Phosphors for Energy Saving and Conversion Technology

Phosphors for Energy Saving and Conversion Technology

Vijay B. Pawade and Sanjay J. Dhoble

CRC Press
Taylor & Francis Group
Boca Raton London New York

CRC Press is an imprint of the
Taylor & Francis Group, an **informa** business

CRC Press
Taylor & Francis Group
6000 Broken Sound Parkway NW, Suite 300
Boca Raton, FL 33487-2742

First issued in paperback 2020

ISBN 13: 978-0-367-57122-1 (pbk)
ISBN 13: 978-1-138-59817-1 (hbk)

Library of Congress Cataloging-in-Publication Data

Names: Pawade, Vijay B., author. | Dhoble, Sanjay J., 1967- author.
Title: Phosphors for energy saving and conversion technology / Vijay B. Pawade and Sanjay J. Dhoble.
Description: First edition. | Boca Raton, FL : CRC Press/Taylor & Francis Group, 2018. | "A CRC title, part of the Taylor & Francis imprint, a member of the Taylor & Francis Group, the academic division of T&F Informa plc." | Includes bibliographical references and index.
Identifiers: LCCN 2018014016| ISBN 9781138598171 (hardback : acid-free paper) | ISBN 9780429486524 (ebook)
Subjects: LCSH: Electroluminescent devices--Materials. | Solar cells--Materials. | Phosphors.
Classification: LCC TK7871.68 .P39 2018 | DDC 621.32--dc23
LC record available at https://lccn.loc.gov/2018014016

Visit the Taylor & Francis Web site at
http://www.taylorandfrancis.com

and the CRC Press Web site at
http://www.crcpress.com

Contents

Part II Phosphors—An Overview

Part III   Roles of Phosphors

Preface

Over the past few years, many universities and colleges have introduced undergraduate and postgraduate courses on green and renewable energy sources and the need for sustainable development considering the adverse effect of globalization on human health and the environment. Therefore, to sustain the environment for the betterment of mankind, it is essential to adopt the concept of "Go green." Currently, this area is gaining more importance and attracting students and researchers due to its necessity in the 21st century. This book provides basic knowledge about the advantages of phosphors in energy saving and conversion devices and the scope of environmentally friendly technology to save energy and the environment. This book is very useful in finding alternative renewable and sustainable energy sources to replace the existing fossil fuel–based polluting technology used in energy generation. The use of clean and environmentally friendly technology really helps to reduce the level of harmful pollutants and greenhouse gases in the environment and to save our lovely planet. This textbook is divided into four parts: the first introduces the fundamentals of atoms and semiconductors, the second deals with an overview of phosphors, the third introduces the scope of phosphors in the development of light-emitting diodes and photovoltaic technology, and the last gives a brief account of the role of energy-efficient technology in sustainable development. Based on these four different parts, the book is further divided into seven chapters. All the chapters were written in a sequential manner and linked to each other for better understanding. The student is encouraged to expand on the topics discussed in the book by reading the references provided at the end of each chapter. Each chapter is written in such a way as to fit the background of renewable and sustainable energy sources and clean technology. The content given in this book can be covered in the undergraduate and postgraduate syllabi of different courses.

Finally, we are grateful to the researchers and publishers who have permitted and provided data wherever necessary to enhance the depth of the book.

About the Authors

Dr. Vijay B. Pawade is an assistant professor in the Applied Physics Department, Laxminarayan Institute of Technology, R.T.M. Nagpur University, Nagpur, India. He obtained his MSc (Physics) in 2008 (specializing in condensed matter physics and materials science) and his PhD (Materials Science) in 2012 from Nagpur University. He has seven years of teaching experience in engineering physics and five years of research experience in the study of optical properties of novel rare earth–doped oxides–based phosphor materials and their application in solid-state lighting technology. He has published 27 research papers in refereed international journals and more than 20 papers presented at conferences. Also, he has contributed four book chapters as a co-author on the characterization of nanomaterials, their engineering applications in energy industries, and their future perspective in the development of environmentally friendly energy sources in *"Nanomaterials for Green Energy"* an edited book published by Elsevier. Now, his research is focusing on the development of lanthanide-based inorganic materials, which find potential applications as multifunctional materials for energy saving solid-state lighting and energy conversion materials for photovoltaic technology. He is the editor of the book entitled *Nanomaterials for Green Energy*, published by Elsevier. He is also a reviewer of many international peer-reviewed journals published by Elsevier, Wiley, Taylor & Francis, the American Chemical Society, and the Royal Society of Chemistry.

Dr. Sanjay J. Dhoble is a professor in the Department of Physics, R.T.M. Nagpur University, Nagpur, India. He obtained his MSc (Physics) in 1988 from Rani Durgavati University, Jabalpur, India, and his PhD in Solid State Physics from Nagpur University, in 1992. He has 27 years of teaching experience and more than 30 years of research experience in the area of luminescent materials and their applications. He has published 570 research papers in international and national journals and has registered two patents. Also, he is the co-author of two international books: *Principles and Applications of Organic Light Emitting Diodes (OLEDs)*, published by Elsevier, and *Phosphate Phosphors for Solid-State Lighting*, published by Springer. He has visited Asian and European countries for research and education purposes. He has successfully supervised 42 PhD students. He is a co-author of 23 books/chapters. He is also an editor of the book entitled *Nanomaterials for Green Energy*, published by Elsevier, and has reviewed many international peer-reviewed papers published by Elsevier, Wiley, Springer, Taylor & Francis, the American Chemical Society, and the Royal Society of Chemistry. Currently, he is an editor of the journal *Luminescence*, published by Wiley.

Part I

Fundamentals of Atoms and Semiconductors

1

Short Review of Atomic and Semiconductor Theory

1.1 Atomic Theory

In the sciences, especially chemistry, physics, materials science, and electronics, atomic theory is a scientific theory that provides basic knowledge about matter, which is composed of tiny units called atoms.

This section gives a short description of the atom. We know that the atom may be considered as the smallest unit of matter that can have different chemical and physical properties [1]. Matter is composed of atoms and molecules. It has four different states: solid, liquid, gas, and plasma, which is the fourth state of matter and consists of neutral and ionized atoms. The typical size of an atom is very small, approximately 100 pm [2, 3]. The atom has no definite size or boundaries, but there are different ways to define the size of an atom. The behavior of a small atom is well understood using the concepts of classical physics. Quantum theory has played an important role in the development of atomic theory and models, and it is also applicable to exploring and predicting the nature of an atom. In basic science, the word *atom* is a philosophical concept from ancient Greece, and it was discovered in the nineteenth century in the field of modern chemistry that matter is indeed composed of atoms. Chemists used the word *atom* in connection with the study of new chemical elements. Later, in the twentieth century, various experiments were carried out based on electromagnetism and radioactivity, for example. Physicists explained that an atom is nothing but a combination of various subatomic particles, which can exist separately from each other. The atomic structure consists of a nucleus at the center, formed by the combination of protons and neutrons, also called *nucleons*, bonded with one or more electrons around the nucleus of the atom. According to the literature, the nucleus makes up 99.94% of the mass of an atom. Protons, which lie inside the nucleus, carry a positive charge, and electrons have a negative charge, such that the atom is electrically neutral. Atoms can be attached to each other by chemical bonds to form a compound. The ability for an atom to associate or dissociate is responsible for the physical changes observed in nature, so it is directly linked with the branch of chemistry.

1.1.1 Short Overview of Atomic Model and Theory

1.1.1.1 Dalton's Atomic Theory

The existence of atoms was first suggested by Democritus, but it took almost two millennia to explain it. However, it gained a foothold with the assumption of an atom as a fundamental object, explained by John Dalton (1766–1844).

Two decades later, Dalton's theory came into existence during the development of modern chemistry. Dalton carried out a most important investigation into the theory of atoms based on chemistry. Hence, the theory was named after Dalton, but its origin was not fully understood [4]. This theory was proposed when Dalton was researching ethylene, methane and analyzing nitrous oxide and NO_2 under the direction of Thomas Thomson [5, 6]. After that he explained the law of multiple proportions based on the idea that the interactions of atoms consist of a chemical combination of definite and characteristic weight [7]. So, while during the study of the properties of atmospheric gases an idea of the atom was in Dalton's mind, it was nothing more than a simple physical concept. He published his idea in 1805. He asked: "Why does not water admit its bulk of every kind of gas alike? This question I have duly considered and though I am not able to satisfy myself completely I am nearly persuaded that the circumstance depends on the weight and number of the ultimate particles of the several gases." Some of the important points discussed by Dalton are as follows:

1. An element is made up of extremely small particles called atoms.
2. They are identical in mass, size, and properties. Also, an atom of a different element has different mass, size, and properties.
3. An atom can be neither created nor destroyed.
4. Chemical compounds are formed by the combination of atoms of different elements.
5. Atoms can be combined, separated or rearranged using a chemical reaction.

1.1.1.2 Rutherford Model

This model was discovered by Ernest Rutherford in 1909. He concluded that the plum pudding model of the atom discovered by J. J. Thomson was not correct. Thus, in 1911, Rutherford worked on the Thomson model [8], and with the help of a gold foil experiment, stated that the atom has a small and heavy nucleus. Later, he designed an experiment in which he used α-particles emitted from a radioactive substance as a probe to investigate the unseen atomic structure. According to this experiment, he predicted that if the beam of emitted α-particles was bombarded and passed in a straight line through the gold foil, then the Thomson model would be correct. But it was found that although most of the radioactive α-rays passed through the gold foil, some of the rays were deflected. Therefore, to give an interpretation of the unexpected experimental result, Rutherford applied his own physical model to the study of the subatomic structure of an atom. In this model, the atom is made up of a central charged nucleus surrounded by an orbiting electron [9]. Rutherford's model deals with charge and the atomic mass of an atom within a very small core but says nothing about the structure of the remaining electron and atomic mass.

The key points of Rutherford's model are as follows:

- An electron moving around the nucleus does not influence the scattering of α particles.
- In many atoms, the positive charge is situated at the center of the atom in a relatively small volume called a *nucleus*. The magnitude of charge and mass are proportional to each other. Therefore, the concentrated central mass and charge of an atom cause the deflection of α and β particles.
- Elements of high atomic mass do not deflect high-speed α particles, which carry high momentum compared with electrons.
- The nucleus is about 10^5 times smaller than the diameter of the atom. This is like putting a grain of sand in the center of a football [10, 11].

1.1.1.3 Bohr Model

In atomic theory, the Rutherford–Bohr model, also called the *Bohr atomic model*, was discovered in 1913. Bohr predicted that the atom has a small, positively charged nucleus and the electrons travel in a circular orbit around it, similarly to a solar system. This is due to the presence of electrostatic and gravitational forces of attraction. The Rutherford model was further improved using the quantum physical interpretation of the atom. The modern mechanical model of the atom has been developed following the Bohr atomic model. According to the laws of mechanics, an electron revolving around the nucleus should release electromagnetic radiation, and due to loss of energy from the electron, finally, it will spiral inward toward the nucleus. Hence, as the orbit becomes smaller and faster, the frequency of radiation should increase with the emission of radiation. Therefore, it would release continuous electromagnetic radiation. However, in the nineteenth century, an experiment was performed with electric discharge showing that atoms emit electromagnetic radiation at certain frequencies.

Therefore in 1913, Bohr proposed a different model, known as *Bohr's model*, to overcome the drawbacks of atomic theory.

Bohr proposed that electrons have classical motion:

1. An electron moves around the nucleus in a circular orbit.
2. An electron has a stationary and stable orbit, without radiating electromagnetic radiation according to its distance from the nucleus [12].

Each orbit has fixed energy, called the *energy level* of an atom. In this case, the acceleration of the electron does not emit radiation, but the energy loss of an electron results from classical electromagnetic radiation.

3. Therefore, an electron in an atom can lose or gain energy when it jumps from a lower to a higher orbit by absorbing or emitting radiation of certain frequencies.

Some key points based on this model are:

1. On the basis of Einstein's theory of photoelectric effect, Bohr's model assumes that during a quantum jump or transition of an atom from one orbit to another, a discrete amount of energy is radiated through it. Therefore, the quantization of the field was discussed on the basis of the discrete energy level of an atom, but Bohr was unable to confirm the existence of photons.

2. Also, Bohr used Maxwell's theory of electromagnetic radiation, according to which the frequency of radiation is equal to the rotational frequency of the electron moving in an orbit, and its harmonics path is an integer multiple of the frequency of an atom. This is obtained from the quantum jump of an electron from one orbit to another on the basis of Bohr's model. These jumps involve the frequency of Kth harmonics in the nth orbit of the electron. For large values of n, the two orbits have nearly the same rotational frequency during the emission process, but the classical orbital frequency of an atom is not clear, whereas for a small quantum number, the frequency does not have a clear classical interpretation. This gives the origin of the correspondence principle, which indicates the requirement of quantum theory in agreement with the classical theory of an atom, which is limited to large n.

3. The Bohr–Kramers–Slater theory violates the law of conservation of energy and momentum during the quantum jump and also it failed to extend the Bohr model

According to Bohr's quantization condition, the angular momentum of an electron is an integer multiple of $\hbar$. It was modified by de Broglie in 1924 as a standing wave condition, where the electron behaves like a wave, and the wavelength of the orbit must fit along the circumference of an electron and is given by

$$n\lambda = 2\pi r$$

By substituting de Broglie's wavelength, $\lambda = h/p$, which reproduces Bohr's rule. In 1913, Bohr justified his rule on the basis of the correspondence principle without giving an interpretation of the wave, in the absence of the wave behavior of particles of matter such as the electron. Again in 1925, quantum mechanics was applied to the modification of atomic theory, in which the electrons revolving in Bohr quantized orbits were studied in detail, and an accurate model of an electron's motion was developed. After that, Werner Heisenberg presented a new theory about the position and momentum of electrons in an orbit. This theory was also based on the concept of wave mechanics, which was discovered by Austrian physicist Erwin Schrodinger. Schrödinger used de Broglie's matter waves to find the solutions of a three-dimensional wave equation. He concluded that electrons would likely move around the nucleus of the atom just as in a hydrogen atom and would be trapped by the positive charge present inside the nucleus.

1.1.1.4 Sommerfeld's Atomic Model

In Bohr's atomic model, Bohr calculated the radii and energies of electrons moving in a stationary orbit around the nucleus; therefore, the observed values were in good agreement with the experimental values. He also studied the spectra of the hydrogen atom.

Therefore, his theory of the atomic model was mostly accepted all over the world. A few years later, Bohr studied the fine spectral lines of a hydrogen atom by using high–resolving power spectroscopes, but unfortunately, he failed to explain it. Hence, considering this failure of Bohr to explain the fine lines of the hydrogen atom, Sommerfeld extended the Bohr study and made some assumptions to improve the atomic theory. Sommerfeld explained that in a stationary orbit, the electrons are not moving in a circular path around the nucleus but travel in an elliptical path. Sommerfeld explained this as being due to the influence of the nucleus located at the center of the atom. The electron moves in an elliptical path around the nucleus as one of its foci, having both a major and a minor axis of its path. During the broadening of the electron's orbit, these two axes are nearly equal in length, and therefore, the electron's path becomes circular. Here, it is also concluded that the circular path is the same as in the case of elliptical path. However, electrons traveling in an elliptical path have angular momentum, and this must be quantized according to the quantum theory of radiations.

Bohr denoted the angular momentum of an electron by using the equation $m = nh/2\Omega$, but Sommerfeld replaced n with an integer k, where the integer k is called an *azimuthal quantum number*. The integers n and k are related to each other as shown by the following:

$$n / k = \text{Ratio of major to minor axis length}$$

At increasing values of k, the path of the electron becomes more eccentric and elliptical. And finally, at $k = n$, the path becomes circular. Thus, when the electron jumps from a higher to a lower level, this jump would be different from those proposed by Bohr, considering that there may be more than one value of k. Thus, Sommerfeld gave a reason for the fine spectral lines of the hydrogen-like atom, and it was shown that the frequencies of some fine spectral lines were in good agreement with the frequencies reported by Sommerfeld. To explain the fine structure of spectral lines, Sommerfeld made two corrections to Bohr's theory:

1. According to the Bohr model, the electron moves in an elliptical path around the nucleus, considering the nucleus as one of its foci.
2. The speed of an electron varies in different parts of its elliptical orbit. This would affect the relativistic mass of an electron moving around the nucleus. Figure 1.1 shows the atomic model proposed by Sommerfeld when elliptical orbits are permitted for electrons, and it deals with two variable quantities: one corresponds to the variation in distance between the electron and the nucleus (r), and the other is due to the angular position of the electron moving around the nucleus, called the *azimuthal angle* (φ).

The main drawbacks of Sommerfeld's model are as follows:

1. It cannot account for the exact number of lines observed in the fine spectra of hydrogen.
2. It is unable to define the distribution and arrangement of electrons in atoms.
3. It fails to discuss the fine spectra of alkali metals (sodium, potassium, etc.).
4. It cannot explain the Zeeman and Stark effect of an atom.
5. It is unable to provide any explanation for the intensities of hydrogen's spectral lines.

1.1.1.5 Free Electron Model

This model introduces the nature of valence electrons in the structure of a metal. In solid-state physics, Arnold Sommerfeld developed the free electron model. During this development, he combined the Drude classical model

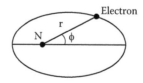

FIGURE 1.1
Sommerfeld atomic model.

with Fermi–Dirac quantum statistics that is called the Drude–Sommerfeld model. The free electron model was successfully applied to study different properties of solids, such as electrical and thermal conductivity, temperature dependence of the heat capacity, density of states, binding energy values, and thermal and electron emission from bulk solids. Therefore, the nature of free electrons moving inside the metal is similar to that of atoms or molecules present in a perfect gas. Thus, the free electrons are also called a *free electron gas*, and the model is based on the free electron gas model. This gas model is different from an ordinary gas in some respects. The free electron gas carries a negative charge, whereas ordinary gas molecules are mostly neutral. Also, the density of electrons in an electron gas is greater than that of molecules in an ordinary gas. Here, the valence electrons or free electrons are also called *conduction electrons,* and they obey Pauli's exclusion principle. Therefore, these valence electrons present inside the metal are responsible for the conduction of electricity. Hence, the free electrons move in a uniform electrostatic field, and the potential energy of the electrons remains constant and is considered as zero. Therefore, the total energy of a conduction electron is equivalent to its kinetic energy. The movement of conduction electrons is strictly restricted within the crystal lattice and the P.E. (potential energy) of a moving electron is less than the P.E. of an identical electron just outside it. Therefore, the energy difference, V_o, of an electron serves as a potential energy barrier that prevents the inner electrons from leaving the surface of the metal. Thus, the motion of free electrons inside a metal is similar to the motion of a free electron gas inside a potential box. Hence, the theory of free electrons can be successfully applied to explain the various properties of metals. The conduction electrons in a metal move randomly and do not carry a current until an electric field is applied to the metal, which accelerates the electrons in a preferred direction. Therefore, the electrons suffer an elastic collision with metal during their motion, which causes a decrease in their speed. Also, metals exhibit high electrical and thermal conductivities, and the ratio of the electrical to thermal conductivity should remain constant for all metals at a constant temperature. This is called the *Wiedmann–Franz law.*

1.2 Basics of Semiconductors

Materials that have conductivity lying between conductors and insulators are called *semiconductors*. They are always in the form of crystalline or amorphous solids having different electrical characteristics [13]. They have high resistance, which is higher than that of other resistive materials but much lower than that of insulators. The resistivity of semiconductors decreases with increasing temperature, meaning that they have opposite behavior to that of a metal. The conducting properties of a semiconductor can be altered by doping an impurity atom into its crystal structure, due to which the resistance of the semiconductor is reduced, and hence, it forms a junction between two different doped regions. At the junction, a current flows due to charge carriers such as electrons, holes, and ions. Semiconductor devices have special features, such as passing current more easily unidirectionally than other devices; also, they have variable resistance and are sensitive to light or heat energy. As discussed earlier, the electrical characteristics of a semiconductor can be improved by doping a suitable impurity ion, and components fabricated from such materials have many advantages in amplification, switching, and energy conversion, for example [14]. The doping impurity in a semiconductor mostly enhances the charge carriers present inside the crystal. If the semiconductor is filled with mostly free holes or positive charge carriers, it is called *p-type*, and if it is filled with mostly free electrons or negative charge carriers, it is known as an *n-type* semiconductor. Thus, the semiconductor device is fabricated by the combination of p- and n-type materials, and the formation of a depletion region between the p- and n-region is responsible for the conduction of electrons and holes from one region to another. Today, most semiconductor materials are widely used in electronic devices and electrical components. Also, some pure elements and compounds have shown excellent semiconductor properties; these include silicon, germanium, gallium, and so on. However, the electrical properties of these semiconductor materials, such as Si, Ge, and GaAs, lie in between those of "conductors" and "insulators." They are neither good conductors nor good insulators; hence, they are called semiconductors. Semiconductor materials contain very few valence electrons, and the atoms are closely packed to form a desired crystalline pattern called a *crystal lattice*, in which the valence electrons are able to move only under certain conditions. Therefore Si and Ge are considered pure or intrinsic semiconductors, because they are chemically pure, and their conductivity is controlled by adding an impurity. The impurity is of two types, donor and acceptor, and by adding it, a number of free electrons or holes can be generated. The properties of semiconductor materials were first studied in the middle of the nineteenth century and also in the first decades of the twentieth century. In 1904, semiconductors were first practically used in electronic devices and integrated circuits (1958) and widely used in early radio receivers [15].

1.2.1 Properties of Semiconductors

Among the different conducting materials, semiconductor materials show excellent electrical and optical properties, which are the unique features of semiconductor devices and circuits. So, these properties can be very well understood by knowing their electrical conductivity, resistivity, thermal conductivity, absorption and emission bands, interband and direct transitions, and so on.

1.2.1.1 Electrical Properties

The electrical properties of semiconductors can be studied on the basis of the resistivity of the materials. For example, metals, such as gold, silver, and copper, have low resistance and are good conductors of electricity, whereas, materials such as rubber, glass, and ceramics carry a high resistance and are unable to conduct electricity. If a semiconductor does not contain any impurities, it cannot conduct an electric current; only by the addition of a suitable impurity can it work as a conductor. Semiconductors made up of only a single compound or element are called *elemental semiconductors* (e.g., silicon). On the other hand, if they are made by the combination of two or more compounds, they are called *compound semiconductors* (such as GaAs, AlGaAs, etc.). Hence, semiconductors constitute a large class of materials having resistivities lying between conductors and insulators. Their resistivity varies over a wide range, and it is reduced by increasing the temperature, as shown in Figure 1.2. Among the different semiconductor elements, silicon (atomic number 14) and germanium (atomic number 32) have the most practical importance, because they have four valence electrons in their outermost shell. Both have a tetrahedral structure, and each atom shares one valence electron with the neighboring atom, forming a covalent bond between them. Therefore, the electrical properties of semiconductor materials reveal that

1. The resistivity of a semiconductor is decreased by varying the temperature; this means that it has a *negative temperature coefficient of resistance*, as shown in Figure 1.2.

    For example, a semiconductor is an insulator at room temperature but becomes a good conductor above room temperature.
2. The resistivity of semiconductors lies in the range 10^{-4} to $10^4 \Omega$-m.
3. The doping of an impurity element such as arsenic, gallium, indium, and so on affects the electrical conductivity of a semiconductor. These have the most important properties from the application point

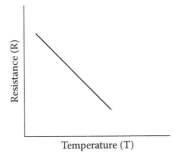

FIGURE 1.2
Temperature dependence of resistance in semiconductors.

of view. Table 1.1 shows the electrical properties of some existing semiconductor materials, including their band gap, conductivity, and so on.

The following list briefly describes some important terms related to the electrical properties of materials.

1.2.1.1.1 Thermal Conductivity

This is the property of a material to transfer a flow of heat from one end to another. It is expressed in the form of the law of heat conduction, also called *Fourier's law of heat conduction*. The rate of heat transfer is lower across a material having low thermal conductivity than one with high thermal conductivity. Materials with high thermal conductivity can be applied as heat sinks, and those with low thermal conductivity are used as thermal insulators in most electrical equipment. Therefore, this property of the materials depends on temperature. Thermal resistivity is defined as the reciprocal of thermal conductivity. In general, metals, such as copper, aluminum, gold, and silver, are good conductors of heat; however, materials such as wood, plastic, rubber, and so on are poor heat conductors, so-called insulators.

1.2.1.1.2 Electrical Conductivity

This is the measure of the capacity of electrical current flow through the material. It is also referred to as *specific conductance*. Put another way, electrical conductivity is defined as the reciprocal of electrical resistivity. The electrical conductivity of the conductor increases as the temperature decreases. In many semiconducting materials, the conduction occurs via charge carriers such as electrons or holes. Therefore, metals (e.g., silver) are materials with good electrical conductivity, whereas insulators, such as glass, pure water, and so on, have poor electrical conductivity. Another example is diamond, which is an electrical insulator due to its regular and periodic arrangement of atoms but is conductive of heat via phonons.

TABLE 1.1

Bandgap and Conductivity of Some Semiconductor Materials

Sr. No	Semiconductor Material	Band Gap (eV)	Conductivity $(\Omega^{-1}\text{-m}^{-1})$
1	Si	1.11	4×10^{-4}
2	Ge	0.67	2.2
3	GaP	2.25	2.2
4	GaAs	1.42	1×10^{-6}
5	InSb	0.17	2×10^{4}
6	CdS	2.40	2×10^{4}
7	ZnTe	2.26	2×10^{4}

1.2.1.1.3 Variable Conductivity

In their normal state, semiconductors act as poor conductors. To start the flow of current in a semiconductor, therefore, requires an electrical or heat supply to its junction, and we know that valence bands (VBs) in semiconductors are filled with free electrons. There are different ways to make the semiconductor a conducting material by doping or gating. The n-type or p-type junction of the semiconductor refers to the shortage of electrons. The current flow through the conductor is due to the unbalanced number of electrons [16].

1.2.1.1.4 Heterojunctions

These types of junction are formed by a combination of two different types of doped semiconducting materials. Generally, they consist of p- and n-doped germanium. The junction shows the movement of charge carriers, such as electrons and holes, between two doped semiconducting materials. The n-doped region of germanium is filled with negatively charged electrons, while its p-doped region would have an excess of positively charged holes. The motion of electrons and holes proceeds by a recombination process, which allows a migration of negatively charged carriers (electrons) from the n-type to combine with the positively charged carriers (holes) from the p-type. Finally, this motion of charge carriers from one region to another results in an electric field [16, 17].

1.2.1.1.5 Excited Electrons

Usually, a greater number of electrons are present in the material when it is in the state of thermal equilibrium. A difference in electric potential would cause a semiconducting material to leave the state of thermal equilibrium and develop a non-equilibrium state. Due to this, electrons and holes interact via ambipolar diffusion. When the thermal equilibrium is disturbed in a semiconducting material, the density of electrons and holes changes. This results in a temperature difference or photons, which enter the system and form electrons and holes. Therefore, this process, which forms and annihilates the present electrons and holes, is referred to as the *generation and recombination process* [16].

1.2.1.1.6 Light Emission

Light emission from semiconductor materials occurs due to the movement of an electron from the lower to the most excited states. Then, it emits light by acquiring a lower energy state without producing a nonradiative transition [18]. Such light-absorbing or -emitting semiconductor materials are used for the fabrication of light-emitting diodes (LEDs) and display devices.

1.2.1.1.7 Thermal Energy Conversion

It is well known that semiconductor materials have large power factors, so they can be applied in thermoelectric systems. This factor depends on the electrical and thermal conductivity of the material, and it is also related to the Seebeck effect and behavior of materials under varying temperatures, and so on [19].

1.2.1.2 Optical Properties

The optical properties of semiconductors involve the interaction of electromagnetic radiation with atoms, which constitutes different processes, such as absorption, emission, reflection, refraction, diffraction, and so on. Electromagnetic radiation is important for the different quantum processes, which shows interesting optical properties of semiconductors.

Optical spectroscopy is an important field in the area of science and technology, which gives us knowledge about the structure and properties of matter (atoms or molecules) based on their spectroscopic characterization [20–25]. Line spectra of atoms were studied in the eighteenth and nineteenth centuries; they provide a great deal of information about the structure and electronic energy levels of the atom [26–28]. Similarly, optical spectroscopy of semiconductors is important in studying the behavior of semiconductor materials after exposure to light radiation. Figure 1.3 shows the band gaps and emission wavelengths of some semiconductor materials. In the early 1950s, detailed knowledge was obtained from semiconductors on the various eigenstates, impurity and defect levels, energy bands, electron excitonic levels, density of states, width of the energy level, symmetries, and so on. The optical properties of semiconductors basically depend on the nature of the electronic band structures and are also related to the lattice structure and bonding between the atoms. Therefore, the lattice symmetry and space groups of an atom are also important to define the structure of energy bands. The optical properties of semiconductors are further classified into electronic and lattice properties, in which electronic properties correspond to the electronic states of the semiconductor, while the lattice properties involve the absorption and creation of phonons during vibrations of the lattice. Thus, the

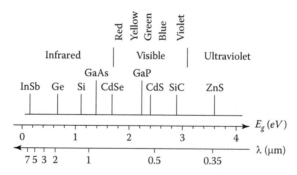

FIGURE 1.3
Band gaps and emission wavelengths of some semiconductor materials. (From http://nptel.ac.in/courses/117102061/LMB2A/3a.htm)

electronic transition corresponding to the optical properties of a semiconductor is called a *one-electron transition*. Therefore, this transition is helpful to find the energy difference between the VB and the conduction band (CB). In the case of the one-electron approximation, each free electron is considered to be a single particle that is freely moving inside a potential well consisting of the combination of core and self-assisted Hartree potential of the other free electrons. This phenomenon is useful in studying the optical properties of materials, such as absorption, reflection, photoconductivity, emission, and light scattering. Today, most of the information about the optical properties can be obtained by photoconductive measurement, but this process is not so simple, due to carriers becoming trapped in the semiconductor; hence, it is difficult to explain. However, the linear optical properties of a semiconductor are referred to as *properties at low light level*. The absorption properties of a semiconductor can be controlled by many physical processes. They depend on the wavelength of the incident light, semiconductor materials, and environmental conditions; that is, pressure and temperature. The electrical and optical properties of a semiconductor can be controlled or modified by adding an impurity to it. The optical properties of pure semiconductors depend on the nature of the materials, whereas the extrinsic properties depend on the doping of impurities, the formation of defects in the semiconductor, and so on [29]. In real solids, many types of defects are present, such as point defects, line defects, planar defects, macroscopic and structural defects, and so on. Thus, the optical behavior of an intrinsic semiconductor can be determined by its lattice vibration, that is, on the basis of the freely available charge carriers, and the corresponding quantum jump of electrons between the different energy states, called *interband transition*. The two different types of electronic transition involved in the semiconductor are described as follows.

1.2.1.2.1 Interband Transition

We know that in semiconductors, the band gap is intermediate between conductors and insulators; that is, it is smaller than in an insulator, such that a sufficient number of charge carriers may be thermally excited at room temperature. Also, sufficient free charge carrier absorption occurs at room temperature through the process of thermal excitation or doping. This type of band transition in semiconductors is generally observed in the infrared and visible region of the spectrum. Hence, the conduction mechanism in semiconductors based on free carriers. If the energy of photons in this process increases and becomes equal to the band gap, there exists a new conduction mechanism, in which incident photons give up their energy to electrons, which excites the electrons from a filled state (VB) to an empty conduction state. So, when a photon is absorbed, an excited electronic level is developed by creating a hole in the VB [30]. This is a quantum mechanical process involving the interaction of radiation with an atom. These interband transitions are further classified as either direct or indirect transitions near the fundamental absorption edge of a semiconductor.

1.2.1.2.1.1 Direct Transitions It is well known that the absorption coefficient of an interband transition depends on the band structure and the photon energy. Direct transitions in semiconductors can be allowed or forbidden based on the dipole matrix element, which gives the transition probability between the energy bands [31]. A nonzero momentum matrix element shows an allowed direction of transitions. In a direct–band gap semiconductor, the momentum of the positively and negatively charged carriers (electrons and holes) is the same in both the CB and the VB, and hence, the electron directly emits light.

1.2.1.2.1.2 Indirect Transitions Here, semiconductors such as GaP, Ge, and Si have maximum energy in the VB and minimum energy in the CB. Therefore, the electron cannot make a direct jump from the VB to CB; in this case, it would violate the conservation of momentum [32–34]. Such a transition or jump is called an *indirect transition*. In the case of an indirect gap semiconductor, an electron cannot emit light, because it passes through an intermediate state by giving the momentum to the crystal lattice.

1.2.2 Energy Bands in Semiconductors

Electrical, optical, and mechanical properties are strongly dependent on the basic arrangement and formation of solids. Different types of bonds exist in solids, such as Van der Waals, metallic bonds, covalent bonds, ionic bonds, and so on. Therefore, to understand these types of bonding, quantum mechanics is necessary to explain the existence of energy levels to move the electrons. This requires knowledge of statistical mechanics to explain the occupation of energy levels by electrons in an atom. Consider two atoms of the same solid being brought close to each other, with the electrons of these two atoms having an identical energy level. When one of the atoms moves to a second position, closer to the other atom, various interactions occur, such as forces of attraction or repulsion between like or unlike atoms. Hence, they balance the interatomic spacing between the atoms. According to Pauli's exclusion principle, these changes result in the electrical properties of a solid. When two atoms are completely isolated from each other, no interaction occurs. As we decrease the spacing between the atoms, they must overlap with each other. The Pauli exclusion principle states that no two electrons can have an identical quantum state; thus, discrete energy level splitting occurs. This splitting takes place for higher energy levels at normal spacing between atoms in a solid. If the distance between atoms approaches the equilibrium interatomic spacing, the energy band splits into two bands separated by an energy gap. The upper band is empty (CB), whereas the lower band is filled (VB). So, the two bands are separated by an energy gap, which contains no allowed energy levels for electrons to occupy. It is referred to as a *forbidden energy gap*. The band gap energy of a material depends on the temperature and interatomic spacing. This is the main characteristic of a semiconductor.

At low temperatures, all energy levels in the VB are filled, and the electrons cannot accelerate to higher energy levels, meaning that no conduction takes place; it occurs only at elevated temperature, because some electrons may have acquired enough energy and jumped to the CB. During this, an empty energy state exists in the VB, and conduction takes place due to positively charged carriers. Thus, the CB consists of free electrons. Therefore, electrical conductivity will be a strong function of temperature. Thus, on the basis of band gaps, metals can be classified as conductors, semiconductors, or insulators. This means that conduction only occurs if there is an electron in the CB and a vacant energy level in the VB. Thus, a semiconductor is regarded as substance that has a completely filled VB and an empty CB separated by a small energy gap of $\cong 1\,eV$. Figure 1.4a and b show the energy band structure of germanium and silicon atoms, respectively. It is seen that the forbidden energy gap of Si is very small, approximately $1.1\,eV$, and for the Ge atom, it is found to be $0.7\,eV$. Thus, a very small amount of energy is required for their valence electron to cross the CB. Therefore, at room temperature, only a few valence electrons are available, and few of them enter into the CB and become free electrons. Hence, at room temperature, a semiconductor element such as germanium or silicon is neither a good conductor nor an insulator. Such a substance is called a *semiconductor*.

Therefore, the relatively small band gap of semiconductors allows an electron to be excited from the lower to the upper band by a certain amount of heat or light energy. At room temperature, a semiconductor with $E_g = 1\,eV$ has a significant number of electrons that are excited thermally across the energy gap into the CB, whereas in the case of an insulator with $E_g = 10.0\,eV$, there is a negligible number of such excitations. The semiconductor band gap ranges from 2.5 to $1\,eV$.

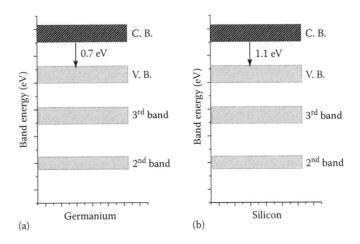

(a)   (b)

FIGURE 1.4
Band gap energy in (a) germanium and (b) silicon.

1.2.3 Effect of Temperature on Materials

As discussed earlier, as the temperature of a semiconductor increases, so does the number of charge carriers (electron or holes), and this increases the conductivity of the material. Semiconductor materials have a negative temperature coefficient, because with increasing temperature the resistance of the material decreases. The following section gives a short description of the effect of temperature on insulators, semiconductors, and conductors.

1.2.3.1 Insulators

In the case of insulators (e.g., glass, plastic, etc.), there exists a large energy ban gap between the VB and the CB. If the temperature of an insulator increases, free electrons go to the upper band. Therefore, in the case of insulators, the resistance decreases with increasing temperature:

$$T \uparrow \qquad R \downarrow$$

1.2.3.2 Semiconductors

Semiconductors act as insulators at room temperature, and the electrons are trapped within the atoms. The conductivity of the semiconductor increases with increase in temperature (i.e., charge carriers in the semiconductor material increase, leading to increased conductivity of the material). Therefore, the outermost valence electrons in the shell acquire energy and jump to another shell of the atom, leaving a hole behind them. Hence, semiconductors have a negative temperature coefficient (i.e., with increasing temperature, the resistance decreases):

$$T \uparrow \qquad R \downarrow$$

1.2.3.3 Conductors

It is well known that the resistance of a conductor depends on the size of the conductor. It is changed by varying the temperature. Thus, the dimension of the conducting material changes; it may either expand or contract. Therefore the change in resistance of the material affects the resistivity of a conducting material, and it also changes the activity of atoms present inside the material. Such a change in the position of atoms may occur due to the flow of charge carriers through the material; that is, the movement of electrons from one region to another under the action of an electric field. If an electrical potential is applied across the two terminals of the conductor, negatively charged electrons are attracted toward the positive potential, and hence, they will migrate from one atom to another. Conductors have a large number of free electrons, and the vibration of atoms in a conductor is caused by collision between the free electrons and captured electrons within their bound state. This blocks the flow of electrons and creates free space. Thus, each collision uses some energy from the valence or free electrons, and this is the main cause of resistance. The value of resistance to the flow of current depends on the number of collisions. Hence, the resistance of the conductor increases with an increase in temperature; that is, conductors have a positive temperature coefficient. For example, Cu^+ is a good conductor of electricity:

$$Cu \qquad Cu^+ + e$$

In the following subsections, we will discuss the effect of temperature on mobility, carrier concentration, and conductivity of semiconductors.

1.2.3.4 Mobility

The conductivity of a semiconductor material is based on two factors: the concentration and the mobility of charge carriers inside it. These two factors depend on the temperature. Therefore, it is important to know the conductivity of a semiconductor as a function of temperature. The mobility of charge carriers inside the semiconductor is influenced by scattering mechanisms. There are two types of scattering: lattice scattering and impurity scattering. In the case of metals, the mobility decreases with increasing temperature, and this is caused by lattice vibrations. In the case of semiconductors, the mobility is affected by the presence of charge carriers. The impurity scattering mechanism is observed due to defects present in a crystal. If the temperature is low, charge carriers travel more slowly, so they have more time to interact with charged impurities. If the temperature of a semiconductor decreases, the impurity scattering increases, and thereby, the mobility decreases. Thus, the total mobility of semiconductor materials is the sum of lattice scattering and impurity scattering.

1.2.3.5 Carrier Concentration

The concentration of charge carriers in a semiconductor is affected by temperature. The intrinsic carrier concentration in a semiconductor is governed by Equation 1.1:

$$n_i = 2\left[\frac{2\Pi kT}{h^2}\right]^{\frac{3}{2}} (m_e\, m_p)^{\frac{3}{4}} \exp\left[\frac{-E_g}{2kT}\right] \tag{1.1}$$

The total carrier concentration is determined by considering the space-charge neutrality. The temperature dependence of electron concentration in a doped semiconductor is shown in Figure 1.5. A small intrinsic electron–hole pair recombination exists at low temperature, where n_i is very small, and the

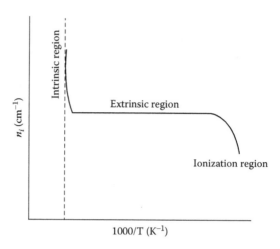

FIGURE 1.5
Plot of carrier concentration (n_i) versus reciprocal temperature for silicon doped with donor atoms.

donor electrons and atoms are bound to each other. This bounded region is called the *ionization region*. If the temperature increases, the electron and atom ionization increases, and all the donor atoms are ionized at about 100 K. At this point, the concentration of carrier is determined by doping. Another region is called the *saturation* or *extrinsic region*, in which available doping atoms are ionized. In this region, it is seen that with increasing temperature, the carrier concentration increases. However, in the case of the intrinsic region, with increasing temperature, the carrier concentration also increases.

1.2.3.6 Conductivity

The conductivity of semiconductor materials depends on carrier concentration and mobility. Also, we know that the conductivity of semiconductors increases with temperature. The conductivity of materials depends on the doping impurity ions, so it is vary with temperature which indicates the dependence of temperature. At high temperature, when the concentration of the intrinsic charge carriers and the mobility of the carriers are dominated by lattice scattering, conductivity can vary with temperature, as shown in Equation 1.2:

$$\sigma \alpha \exp\left(\frac{-E_g}{2kT}\right) \tag{1.2}$$

where:
- σ is conductivity
- E_g is the energy gap
- K is the Boltzman constant
- T is temperature

This equation indicates that the conductivity of the semiconductor depends on the band gap and temperature.

1.2.4 Hole Current

We know that a hole is a positive charge carrier, and it is equal in magnitude but opposite in polarity to an electron. These two charge carriers, holes and electrons, are responsible for the flow of current in semiconductor materials.

In the case of a pure semiconductor at room temperature, a few of the covalent bonds will break, and this will generate free electrons. When an electric field is applied to the material, these free or valence electrons constitute a flow of electric current. At the same time, there is another, positive current carrier, called the *hole current flow*, in the opposite direction to the flow of electrons. Holes are created when the covalent bonds in a semiconductor break due to thermal energy, which removes one electron and leaves a vacancy due to the lacking electron in the covalent bond; this vacant space is called a *hole* or *positive charge carrier*. Thus, releasing one electron from the VB will create one hole. Therefore, electron–hole pairs are created easily by thermal energy [35]. Figures 1.6 and 1.7 explain the mechanism of current conduction by holes. As shown in Figure 1.6, suppose the valence electron at A has been released due to thermal energy; this creates a hole in A. The hole is a prominent center of attraction for electrons due to their opposite charges. Now, the free electron at B will become available from the nearby covalent bond with A and fill the vacant space at A. This creates a vacancy at B. Another free or valence electron, say at C, in turn may leave its covalent bond to fill the vacant space at B, creating a hole at C. Therefore, in this case, the hole, having a positive charge, has been moved from A to C; that is, toward the negative end terminal of the source. Thus, finally, this results in the hole current. Hence, it is noticed that the hole current is only observed due to the moving of valence electrons from one covalent bond to another. Therefore, this phenomenon can be discussed here on the basis of energy band structure, as shown in Figure 1.7. Therefore due to thermal energy, an electron leaves the VB and occupies the empty CB, as shown in Figure 1.7. This will leave a hole at A. Now, another free electron

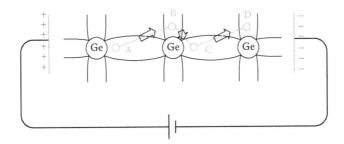

FIGURE 1.6
Path of electrons moving along the direction of ABCD.

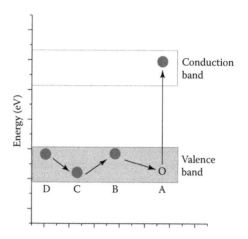

FIGURE 1.7
Path of holes moving in the opposite direction (DCBA).

at B comes to fill the hole at A. The result is that the hole at A disappears due to filling by a valence electron, and a hole appears at B. Another valence electron at C moves into the hole created at B. At the same time, a hole is created at C. Therefore, it is seen that valence electrons move along the path ABCD, whereas holes move along the opposite path, DCBA.

1.2.5 Types of Semiconductor

1.2.5.1 Intrinsic Semiconductor

This is also called a *pure* or *undoped semiconductor*, without the addition of any type of impurity ions. The properties of a material determine the number of charge carriers present in it, besides the amount of impurities. In the intrinsic type of semiconductor, the numbers of negatively and positively charged carriers are equal ($n = p$); that is, the number of electrons in the VB is equal to the number of holes in the CB. If this type of semiconductor is doped with equal amounts of donor and acceptor impurities, it is intrinsic in nature. Therefore, it is not an *undoped semiconductor*. Also, electrical conductivity of intrinsic semiconductors is observed due to the presence of defects (vacancy, interstitial, and antisite defects, etc.) or electron excitation, and so on. In this semiconductor, the maximum energy of the VB occurs at a different momentum-like vector—also referred to as a *wave vector*—associated with electrons in a crystal lattice than the minimum energy of the empty CB. Silicon and germanium are well-known examples of this type of semiconductor. Further, a direct band gap semiconductor is one in which the maximum energy of the VB is at the same value of wave vector as that of the CB, which is at the minimum energy. GaAs is a suitable example of such a kind of semiconductor. Normally, the silicon crystal behaves like an insulator at room temperature, but at a temperature above absolute zero, there is a finite probability that an electron loosely bound in its lattice shell will be knocked easily from its position, leaving a hole in its shell. If we apply a voltage, the movement of electrons and holes shows a small current flow through a crystal. When a semiconductor is kept at ordinary temperature, there is a finite possibility that electrons will jump to the CB and become involved in an electrical conduction process. When silicon is at a temperature above absolute zero, there will be some electrons in the VB that are excited and jump to the CB, supporting the flow of charge.

1.2.5.2 Extrinsic Semiconductor

An extrinsic semiconductor is also called a *doped semiconductor,* to which an impurity has been added to vary the properties of the semiconductor from those of the pure semiconductor. The doping changes the electron and hole carrier concentrations of the semiconductor at thermal equilibrium. Thus, extrinsic semiconductors, such as n-type or p-type, are further classified on the basis of their dominant carrier concentrations. Therefore, the electrical properties of extrinsic semiconductors are the most important parameters to show the advantages of semiconductors in many electronic devices. So, doping can change a semiconductor's properties from intrinsic to extrinsic. The impurity atoms used for doping are atoms of a different element than those elements that are used in intrinsic semiconductors, as given in the periodic table. Therefore, the atom or element used for doping is of either a donor or an acceptor type, which, on doping an intrinsic semiconductor, changes the electron and hole concentrations in the semiconductor. Donor atoms have more valence electrons than the intrinsic semiconductor lattice, where they occupy the state and replace it. The doping atoms give their outermost unfilled valence electrons to another conduction level or band by giving excess electrons to the pure semiconductor. This excess number of electrons then increases the concentration of charge carriers (n_0) of the semiconductor by making them n-type. Another type of semiconductor is the p-type, in which the acceptor atoms have fewer free electrons in the outer shell than the semiconductor atoms, and hence, they replace the pure semiconductor lattice by "accepting" electrons from the semiconductor's VB, thereby providing an excess of positively charged carriers to an intrinsic semiconductor. Thus, the large number of holes increases the hole carrier concentration (p_0) of the semiconductor, making it a p-type semiconductor. In the periodic table, the semiconductor elements from group IV use group V and group III elements as donor and acceptor atoms. In the class of group III–V semiconductors, elements belonging to groups II and VI are used as donor and acceptor atoms. Semiconductor elements in groups III–V use group IV atoms as donors or acceptors. When an element of group IV replaces a group III element in the semiconductor lattice, the group IV element acts as a donor. Conversely, when a group IV replaces a group V element, the group IV element acts as an acceptor. Therefore, group IV elements can be used for both purposes, as donors as well as acceptors, and hence, they are called *amphoteric impurities*.

1.2.5.2.1 Uses of Extrinsic Semiconductors

An extrinsic or doped semiconductor is a key component of many electronic devices. A unipolar semiconductor device (e.g., a diode) is formed by the combination of p-type and n-type semiconductors. Currently, most semiconductor diodes are made up of silicon or germanium materials. Transistors are made up of extrinsic semiconductors; the most commonly used transistors are bipolar junction transistors (BJTs), NPN, PNP, and NPN. Other electronic devices that are fabricated using extrinsic semiconductors include

- Lasers
- Solar cells
- Photodetectors
- Light-emitting diodes
- Thyristors

1.2.5.2.2 Doping

As discussed in Section 1.2.5.2, doping is an important term in semiconductor materials, which introduces impurities into a pure intrinsic semiconductor to modify the related properties of the materials. If the semiconductor is lightly doped, it acts as an extrinsic semiconductor, whereas if the semiconductor is heavily doped like a conductor, it is called a *degenerate semiconductor*.

1.2.5.2.3 Effect of Doping on Band Structure

In the case of light-emitting materials such as phosphors and scintillators, the doping impurity is known as an *activator*. When an impurity is doped into the semiconductor crystal structure, it forms allowed and possible energy levels within the semiconductor. Therefore, the energy levels that are formed by the dopant inside the material are close to the energy band levels of the semiconductor. When a donor impurity is doped into a semiconductor, it forms the electron donor level near the CB, whereas acceptor impurities form the acceptor level near the VB. The energy gap between the energy level formed and the nearest energy band level is called the *dopant site energy band* (E_B), and it is relatively small in value (E_B for boron in silicon = 0.045 eV, and Si = 1.12 eV). The E_B for the doping atom is so small that the atoms are thermally ionized at room temperature, creating free charge carriers in the VB or the CB, respectively. The doping atom may also affect the shifting of energy level with respect to the Fermi energy level.

Acceptor and Donor Impurities Used in Silicon

1. *Acceptors*: Boron, aluminum, nitrogen, gallium, indium [36, 37].
2. *Donors*: Phosphorus, arsenic, antimony, bismuth, lithium [38–41].

1.2.5.2.4 P-Type Semiconductors

P-type semiconductors are formed by doping a trivalent impurity atom, such as boron, aluminum, or gallium, into a pure semiconductor, which creates a deficiency of holes. P-type semiconductors materials consist of a majority of positively charged carriers (hole) carrier concentration. The bonding and structure of p-type semiconductors are shown in Figure 1.8; they are formed by doping a pure semiconductor with acceptor impurities. A commonly used p-type dopant for silicon is boron. In case of p-type semiconductors the Fermi level lies below the intrinsic level but it found closer to the VB than the empty CB.

1.2.5.2.5 N-Type Semiconductors

N-type semiconductors are filled with the maximum negatively charged carrier (electron) concentration. Such a semiconductor is formed by adding an intrinsic semiconductor with donor impurities, as listed earlier. Antimony is the most commonly used dopant for n-type silicon. Figure 1.8 shows the bonding and structure of an n-type semiconductor. The Fermi level is located at a higher level than that of an intrinsic semiconductor and is found closer to the CB than the VB.

1.2.6 Wide Band Gap Semiconductor

A semiconductor is also referred to as a *small band gap insulator*. The doping impurity atoms alter the electronic properties of the semiconductor. There are different applications of such materials in computer and photovoltaic (PV) technology; also, they are used in many electronic devices, as discussed in the previous section. Semiconductors are a basic and emerging field in the development of electronic materials technology. Most semiconductor materials are crystalline in nature; this means that they are inorganic materials, based on the constituent atoms present inside them. Thus, different semiconductor

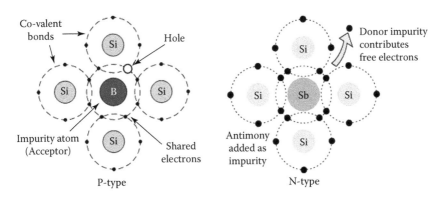

FIGURE 1.8
Bonding and structure of p- and n-type semiconductor.

materials have different electrical and optical properties [41]. In LEDs, the band gap of the material determines the emission wavelength range; if the material has a large band gap, emission occurs toward the shorter wavelength end, whereas if the material has a smaller band gap, the emission is always shifted toward the longer wavelength end. The band gap value of a GaAs semiconductor is 1.4 eV, and this amount of energy corresponds to a wavelength of approximately 890 nm, in the near infrared region. Therefore, a GaAs-based PV cell is unable to convert a light of shorter wavelength into electric current. Silicon has a band gap energy of the order of 1.1 eV, which corresponds to a wavelength of 1100 nm—this is even worse. The band gap required for a single-junction PV cell to perform solar energy conversion has been reported at around 1.0–1.5 eV [42]. Thus, larger–band gap materials have the ability to capture or absorb nearly the entire wavelength of the incident solar radiation that reaches the earth's surface, but a smaller–band gap material, in the case of a single-junction cell, cannot absorb a large portion of light radiation, meaning that it is unable to convert the shorter-wavelength light of the incident solar radiation. Therefore, considering this drawback in solar cell devices and technology, today there is an urgent demand for advanced wide–band gap materials. Currently, researchers are busy developing multi-junction solar cells that can absorb some parts of the incident solar radiation more efficiently. For this purpose, only wide–band gap PV materials have the ability to collect the part of the spectrum beyond the infrared region. LEDs are also fabricated using wide–band gap nitride semiconductors. The relation between the absorption wavelength and the band gap energy in semiconductors is that it is the minimum energy required to excite an electron into the CB. To excite an unassisted photon, more energy is required. But when the excited electron–hole carriers undergo recombination, photons are generated, having the energy that corresponds to the magnitude of the material's band gap. In the case of indirect gap semiconductors, there are not enough photons to achieve the mechanism of absorption or emission, and hence, such indirect–band gap semiconductors are usually very inefficient emitters, although they work better as absorbers. Silicon and some other semiconducting materials have a band gap varying from 1 to 1.5 (eV), which suggests that such semiconductor devices can be operated at relatively low voltages. However, the wide–band gap semiconductors, which have band gaps observed in the range of 2–4 eV, work at much higher temperatures, of the order of 300 °C. This makes them highly interesting for military applications. There follows a list of some wide–band gap materials used in semiconductor devices.

Some large–band gap semiconductors

- SiC.
- SiO_2.
- AlN.
- GaN.

- Boron nitride, h-BN (hexagonal form), and c-BN (cubic form) can form ultraviolet (UV) LEDs.
- Diamond.

1.3 Germanium and Silicon Atoms

1.3.1 Germanium (Ge)

Germanium has the atomic number 32, and its orbit consists of [2, 4, 8, 18] electrons, respectively. It is quite similar to tin and silicon in some respects. So, this metal acts as a pure semiconductor. It generally reacts and forms complexes with oxygen in nature. It is a reactive element and found everywhere on the earth's surface. It is a popular semiconductor, which has been used over the past few decades in the fabrication of transistors and various other electronic devices. Nowadays, the advantages of this semiconductor are increasing due to its extensive properties. It is applicable in the fields of fiber-optic technology, infrared optics, solar cells, LEDs, and so on. It is also used in the production of nanowires. At the beginning of the development of semiconductor technology, many types of electrical components were made using germanium, but later, it was observed that they spontaneously extruded very long screw dislocations in the materials, which is one the most important reasons for the failure of semiconductor components made from germanium. Recently, Si-based solar panels have been fabricated using germanium, because it acts as the substrate of the Si wafers for high-efficiency multi-junction PV cells. It also acts as a better reflector, used for automobile headlights and to backlight liquid crystal display (LCD) screens; it plays an important role in high-brightness LEDs [43]. The lattice parameters of Ge and GaAs are very similar; therefore, Ge substrates are applicable for making GaAs-based solar cells [44]. Also, filaments made from Ge are used in fluorescent lamps and solid-state light-emitting technology [45]. The orbital structure of Ge belongs to the family of carbon and silicon. All these elements have four valence electrons in the outermost shell.

1.3.2 Silicon (Si)

Si is a tetravalent metalloid having the atomic number 14. It belongs to group 14 in the periodic table. Elemental silicon is easily found everywhere in the earth's crust, and it has many application, so it has a large impact on the modern world economy. Many types of integrated circuits (ICs), computer parts, cell phones, and modern technology equipment are fabricated using Si. Only a small amount of very highly purified silicon is used to fabricate ICs. There are two types of silicon used in electronic devices: *amorphous* and *polycrystalline silicon*.

1.3.2.1 Amorphous Silicon

This is not in a very crystalline form and is therefore called *amorphous*. It is used for the fabrication of solar cells and display devices. It is always applicable for depositing on a variety of flexible substrates, such as glass, metal, plastic, and so on. Solar cells made from the amorphous silicon generally have low light conversion efficiency, but they are the most environmentally friendly PV cells, because they do not contain any toxic heavy metals such as Cd or Pb. Thus, amorphous silicon has shared the maximum contribution in the fast-growing PV market and has become a leading source of second-generation thin-film solar cell technology. But today, its significance is decreasing due to strong competition from the currently used crystalline silicon cells and other thin-film technologies, which have better light conversion efficiency. Amorphous silicon differs from monocrystalline silicon, a single crystal, and polycrystalline silicon, which consists of small grains, also called *crystallites*.

1.3.2.2 Polycrystalline Silicon

Also called *poly-Si*, this is a highly purified form of silicon used for the fabrication of solar PV and electronic devices and components. Generally, polysilicon contains the lowest impurity levels—less than 1 ppb—whereas polycrystalline solar grade silicon is relatively less pure [46]. It is used in the form of slices in silicon wafers for the manufacturing of solar cells, ICs and other semiconductor devices, and so on. This polysilicon consists of small crystallites, while polycrystalline silicon and multisilicon are multicrystalline, usually referring to crystallite size larger than 1 mm. Therefore, multicrystalline solar cells are the most common type of solar cells used in PV technology. As per the literature report, the production of 1 megawatt conventional solar modules requires 5 tons of polysilicon [47]. Hence, polysilicon is different from monocrystalline and amorphous silicon.

1.3.2.3 Monocrystalline Silicon

This is also referred to as *single-crystal silicon* or *mono-Si*. Si is the base material, which is cheap and has been used to fabricate many electrical components for the last few decades. It is used in PV panels as a light-absorbing material. Due to this major advantage, it has gained importance in silicon technology. Nowadays, it is available everywhere at low cost and applicable for the fabrication of many silicon-based electronic devices. The monocrystalline form of silicon is different from other allotropic forms, such as non-crystalline or amorphous and polycrystalline silicon. The efficiency of monocrystalline (Si) cells is higher than that of polysilicon cells (20.4%), copper indium gallium diselenide (CIGS) cells (19.8%), CdTe thin-film cells

(19.6%), and amorphous Si cells (13.4%) [48], in the commercial PV market. Currently, the fabrication and processing costs of silicon thin films are decreasing significantly due to improvements in technology, and the device cost of silicon-based solar cells has fallen to some extent. Currently, enhancements in the efficiency of the devices and the use of low-cost and easily available materials for the development of technology are the key ideas for reducing the cost of the devices [49].

1.4 Solar Energy and Solar Cells

Solar energy is the light energy emitted by the sun, which is used in different technologies such as solar heaters, PV, solar thermal energy, solar architecture, and so on [50]. Hence, it is an evergreen source for the development of renewable energy and its related technologies. These are broadly classified as passive or active devices, depending on how this technology is useful to capture and convert incident light into solar power output. Active solar technology based on PV systems uses concentrated solar power and solar water heaters to harness the solar energy, whereas passive technology is based on PV solar film, which may be spread on the windows, walls, roof, and floors of buildings to capture the maximum solar radiation incident on it and convert it into energy. Recently, solar energy has become an ideal: the most promising, clean, and pollution-free energy source for the twenty-first century [51]. It is also called a *universal natural source*, an *infinite reserve*, and it has the advantage of being a clean and economical source for sustainable development. Due to its scope, solar energy has been widely adopted as a renewable energy source in many countries [52, 53]. In 2011, the International Energy Agency reported that "the development of affordable, inexhaustible and clean solar energy technologies will have huge longer-term benefits." So, the use of solar energy will fulfill the need for energy security, help to sustain the economy, reduce pollution, reduce global warming, and reduce fossil fuel prices. These are a few of the global advantages of solar energy.

1.4.1 Types of Solar Cell

The types of solar cell that are fabricated using silicon are as follows. Table 1.2 shows the existing cell efficiency of the silicon solar cell.

- Crystalline silicon (c-Si) solar cells
- Perovskite solar cells
- Polycrystalline silicon

TABLE 1.2

Different Solar Cell Materials and Their
Conversion Efficiency (%)

Material	Efficiency (%)
Monocrystalline silicon	14–17
Polycrystalline silicon	13–15
Amorphous silicon	5–7

1.4.2 Basics of Solar Cells

1.4.2.1 Construction and Operation

In general, solar cells are composed of doped silicon and metal contacts, as shown in Figure 1.9. The doping is done as a thin layer spread on the top of the cell, which forms a p–n junction with a band gap energy E_g. The junction diode is made of Si or GaAs semiconductor materials, in which a p-type thin layer is deposited on the n-type, while the top of the p-layer is attached to a fine metal contact, which provides the space for light to reach the thin p-layer. The current-collecting electrode is provided at the bottom of the n-layer. Figure 1.9 shows the conduction of charge carriers (i.e., electrons and holes in the p–n region) generating the photocurrent through the electron–hole recombination process. In solar cells, electrons and holes are accelerated toward the p-doped and n-doped region by the scattering electric field E_{scatt} in the quasi-neutral region, which generates a scattering photocurrent I_{pscatt} (I_{nscatt}). Therefore, due to the accumulation of charge carriers, a voltage (V) and a photocurrent (I_{ph}) are produced under sunlight illumination. In the quasi-neutral region, the photons give up their energy to electrons. Hence, electrons move from VB to CB due to the electric field, thereby generating a photocurrent as shown by Equation 1.3:

$$I_{ph} = I_g + I_{nscatt} + I_{pscatt} \tag{1.3}$$

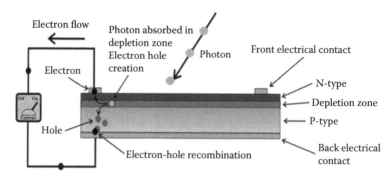

FIGURE 1.9
Construction of solar cell device.

where:

I_g is the generated current

The current-voltage (J-V) characteristics of a solar cell (i.e., current per unit area of a cell) are obtained under illumination by shifting the J-V characteristics of a diode in the dark by downward photocurrent. Therefore, solar cells are fabricated specially to supply power only, and the power is related to the equation $P = V \cdot I_{ph}$, which must be negative. To obtain the maximum power, the starting points (V_m, J_m) have to be located in the region between $V > 0$ and $I_{ph} < 0$, respectively (Figure 1.10).

1.4.2.2 Loss Mechanisms

The theoretical cell response of the Si solar cell was first reported in the 1960s, and it is called the *Shockley–Queisser* limit. This limit exhibits several loss mechanisms that affect solar cell design. The first loss is attributed to blackbody radiation, and it accounts for about 7% of the power at standard temperature and pressure. The second loss is observed due to the recombination effect and is of the order of 10%. Thus, the dominant loss mechanism in a solar cell is defined as the inability of a solar cell to remove all the power in the photon, and the associated problem is that it cannot remove any power at all from certain photons. Figure 1.11 shows a plot of band gap versus efficiency of the solar cell. Photons have enough energy as compared to the band gap of the solar cell. If the photon carries lower energy than the band gap of the cell, then the cell is unable to collect the power. Generally, such effects are observed in conventional solar cells, which are not sensitive

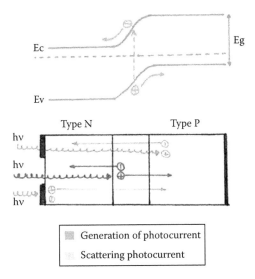

FIGURE 1.10
Generation of photocurrent through a solar cell.

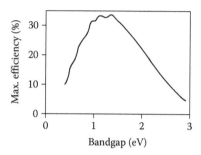

FIGURE 1.11
Graph of band gap versus efficiency of a single-junction solar cell.

to the infrared (IR) region of the spectrum. Photons with higher energy than the band gap of the cell, which corresponds to blue light, initially eject an electron to a higher state above the band gap, and this excess energy of the photon is lost through collisions called the *relaxation process*. This lost energy produces heat in the cell and thereby, increases the blackbody losses [56]. Besides this loss, the maximum efficiency for a silicon cell is found to be about 34%. This means that approximately 66% of the sunlight energy will be lost on hitting the cell.

1.5 PV Solar Cells

A PV cell is also called a *solar cell*, which converts light energy directly into electricity based on the principle of the PV effect. When light rays are incident on a PV cell, due to quantum mechanical processes, some part of the light is reflected, absorbed by the atoms. Thus, the conversion of photocurrent is a physical phenomenon. In a PV solar cell, the electrical characteristics, such as current, voltage, or resistance, vary with respect to incident light radiation.

The operation of a PV cell is based on the following factors:

- Absorption of incident sunlight to generate either electron–hole pairs or excitons
- Separation of charge carriers of opposite sides
- Separate extraction of charge carriers to an external circuit

PV cells are mostly manufactured using a semiconductor material such as Si. In general, it is the combination of metals and insulators, which are selected on the basis of their properties. These materials have the capacity to convert light energy into electricity. The amount of electric current generated by the PV cells depends on the quality of incident light and the output V-I characteristics of the cell. Today, different types of PV cells are available in the market, such as silicon PV, thin film PV, organic PV, or concentrator photovoltaics (CPV) solar cells. The rate of photoelectron emission in the PV materials will increase in proportion to the intensity of incident light.

The light absorbed by a solar cell is the combination of directly incident solar radiation and the amount of diffuse light that is bounced off the surrounding surfaces. An anti-reflective (AR) coating is spread on the solar cell module, which is helpful to enhance the light absorption capacity of the cell. Generally, a material such as silicon nitride or titanium oxide is used for the AR coating. The PV cells can be constructed in different configurations to form a module, cell, and array, as shown in Figure 1.12.

In 1954, Bell Laboratory invented the first practical solar cell. Due to long and continuous discoveries and research developments in the field of solar science and technology, today, new, advanced solar cells have led to tremendous growth in the PV market [54–60]. The current-voltage characteristics and some important parameters related to PV cells are discussed in the following section.

1.5.1 I-V Characterization of PV-Solar Cells

Figure 1.13 shows the I-V characteristics of a PV solar cell under light illumination, and the direction of conventional current and voltage is also indicated in the corresponding circuit diagram, just like a semiconductor diode. In the absence of light illumination, the PV cell behaves like a diode, and it cannot generate any photocurrent. Thus, the production of photocurrent in the PV cell depends on the intensity of incident light.

Figure 1.14 shows the equivalent circuit of a PV solar cell. It consists of different components, such as incident light source or current source (I_l), diode, series resistance (R_s), and shunt resistance (R_{SH}). Series resistance (R_s) shows the ohmic losses observed in the front surface of the cell, and shunt resistance (R_{SH}) indicates the loss observed due to diode leakage current [61–64]. Therefore, the conversion efficiency (η) of the PV cell is defined as the ratio of

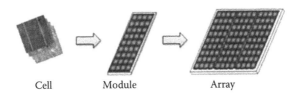

Cell   Module   Array

FIGURE 1.12
Solar cell, model, and array.

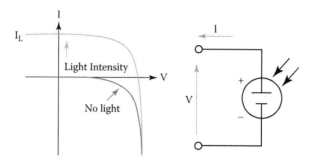

FIGURE 1.13
I-V curve and circuit diagram of PV cell. (From www.ni.com/white-paper/7229/en/ [65].)

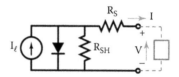

FIGURE 1.14
Diode-resistance equivalent circuit for a PV cell. (From www.ni.com/white-paper/7229/en/.)

maximum electrical power that the array or cell can produce to the amount of solar irradiance hitting the array and is given by Equation 1.4:

$$\eta = \frac{P_{max}}{P_{in}} \tag{1.4}$$

where:
P_{max} is the max power output

The corresponding fill factor (i.e., the relation between maximum power output of the array under normal operating conditions and the product of the open-circuit voltage and the short-circuit current) (FF) is given by

$$FF = \frac{I_m V_m}{I_{SC} V_{OC}} \tag{1.5}$$

where:
P_{in} is the input power of the cell
V_{OC} is the open-circuit voltage
I_{sc} is the short-circuit current
I_m is the maximum cell current
V_m is the maximum voltage of the cell

In the case of an ideal PV cell, the total current I through the cell is equal to the difference between the current ($I\ell$) generated by the principle of photoelectric effect and the diode current (I_D). It can be represented by the Equation 1.6:

$$I = I_\ell - I_D = I_\ell - I_0\left(e^{\frac{qV}{KT}} - 1\right) \tag{1.6}$$

where:

I_0 is the saturation current of the diode
q is the charge on the electron (1.6×10^{-19} Coulombs)
k is a constant (1.38×10^{-23} J/K)
T is the cell temperature in Kelvin
V is the cell voltage

Therefore, the accurate model of PV cells consists of two diode terms, but here, this will be discussed using a single-diode model. Thus, based on Equation 1.6, the simplified circuit model is shown in Figure 1.3, and the related equation for current is given in Equation 1.7:

$$I = I_\ell - I_0\left(\exp^{\frac{q(V + I R_s)}{n K T}} - 1\right) - \frac{V + IR_s}{R_{SH}} \tag{1.7}$$

where:

n is the diode ideality factor
R_S and R_{SH} are the series and shunt resistance, respectively

The corresponding output I-V curve of the PV cell is shown in Figure 1.15, in which the voltage changes from zero to V_{OC}, and the corresponding PV cell parameters, such as the equivalent cell shunt and series resistance, electrical conversion efficiency, and fill factor, can also be determined from I-V measurements, as reported in the next section. Here, a light source with a constant intensity and a known spectral distribution must be used, and the corresponding cell is kept at a constant temperature.

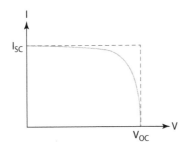

FIGURE 1.15
Illuminated I-V sweep curve of PV cells. (From www.ni.com/white-paper/7229/en/.)

Some important parameters related to the I-V study of PV solar cells are given in the following subsections.

1.5.1.1 Short-Circuit Current (I_{SC})

The short-circuit current (I_{SC}) corresponds to the short-circuit condition in which the resistance is kept at a low value and the voltage must be zero:

$$I_{(V=0)} = I_{SC} \tag{1.8}$$

I_{SC} corresponds to the maximum current value for forward-bias sweep in the power quadrant. In the case of an ideal cell, this maximum current (I_{max}) is the total current produced in the solar cell after the photon excitation, and it is denoted by Equation 1.9:

$$I_{max} = I_{SC} = I_\ell \tag{1.9}$$

1.5.1.2 Open-Circuit Voltage (V_{OC})

This occurs when there is no current passing through the PV cell.

$$\therefore V \,(\text{at } I = 0) = V_{OC} \tag{1.10}$$

where V_{OC} is the maximum voltage difference across the cell for a forward-bias sweep.

$$\therefore V_{OC} = V_{max} \tag{1.11}$$

1.5.1.3 Maximum Power (P_{max})

The output power of the PV cell (Watts) can be calculated from the I-V curve, as shown in Figure 1.16, and it is denoted by Equation 1.12:

$$P = IV \tag{1.12}$$

From the curve, it is seen that at the I_{SC} and V_{OC} on the I-V curve, the power will be zero, and the maximum power will be observed between these two points, indicated by the dark spot on the curve. The corresponding values of voltage and current at this maximum power point are denoted by V_{MP} and I_{MP}, respectively.

1.5.1.4 Fill Factor (FF)

This is an important parameter to study the quality of the solar cell. It is estimated by comparing the ratio of maximum power with the theoretical

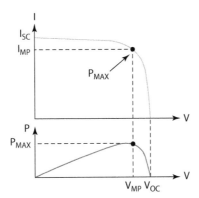

FIGURE 1.16
I-V curve showing maximum power of a PV cell. (From www.ni.com/white-paper/7229/en/.)

power (P_T) of the cell. Hence, it is the product of output open-circuit voltage and short-circuit current. Therefore, the fill factor can be represented by the rectangular areas shown in Figure 2.7. A larger fill factor is desirable if the corresponding I-V sweep looks like a square. For the PV cell, the fill factors are observed in the range from 0.5 to 0.82 (Figure 1.17).

1.5.1.5 Efficiency (η)

The efficiency of a PV cell is defined as the ratio of the power output (P_{out}) to the power input (P_{in}). Here, the output power is denoted by P_{max}; hence, the solar cell can operate up to its maximum power output to get the maximum efficiency, as represented by Equation 1.13:

$$\eta = \frac{P_{out}}{P_{in}} \Rightarrow \eta_{max} = \frac{P_{max}}{P_{in}} \tag{1.13}$$

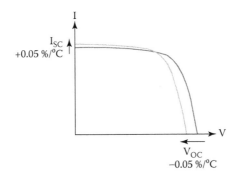

FIGURE 1.17
Graphical representation of the fill factor from the I-V sweep curve. (From *Green, M.A.*, Third Generation Photovoltaics, *Springer-Verlag, 2003.*)

where P_{in} is the product of current and voltage generated during the irradiance of the incident light, measured in Watts per square meter with respect to the surface area of the solar cell or module (square meters). The standard laboratory test of PV devices is carried out with a constant light source, and the temperature of the surrounding atmosphere is kept constant to avoid any errors during the study.

1.5.1.6 Temperature Measurement Considerations

PV cells are made from those semiconductor materials that are temperature sensitive. Figure 1.18 shows the effect of temperature on the I-V characteristics of the cell. When a PV cell is exposed to high temperature, this affects the I-V curve, in which I_{SC} increases slightly; however, V_{OC} decreases more significantly.

The increase in temperature of the cell may also affect the maximum power output of the cell. The temperature of the device can be measured with the help of sensors such as resistance temperature detectors (RTDs), thermistors, thermocouples, and so on.

1.5.1.7 I-V Curves for Modules

In a PV module, the nature of the characteristic I-V curve does not change, but its scale is based on the number of cells connected in series and parallel position. If n is the number of cells connected in series, and m is the number of cells connected in parallel, and I_{SC} and V_{OC} are corresponding parameters for the module, the resulting I-V curve is as shown in Figure 1.19.

1.5.2 Advantages of Si Solar Cells

The c-Si solar cell can give better efficiency: up to 27.6%. It shows little degradation over time. Therefore, most solar cells are made using crystalline silicon. It is well known that mono-c-Si cells belong to the first generation

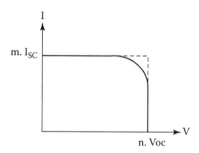

FIGURE 1.18
Temperature effect on I-V curve. (From www.ni.com/white-paper/7229/en/.)

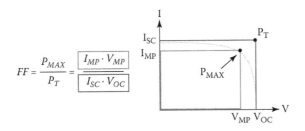

$$FF = \frac{P_{MAX}}{P_T} = \boxed{\frac{I_{MP} \cdot V_{MP}}{I_{SC} \cdot V_{OC}}}$$

FIGURE 1.19
I-V curve for PV module. (From www.ni.com/white-paper/7229/en/.)

of PV technology and have a long lifespan. This type of cell is formed from high-purity materials, and their structure is perfect, which makes for very efficient energy conversion. The effectiveness of the cells is due to lack of grain boundaries, so that in the case of silicon, there is no grain boundary to obstruct the movement of electrons around the silicon structure. Therefore, silicon solar cells have some good advantages and few disadvantages. The efficiency of monocrystalline solar cells is higher, but the cost is also higher; polycrystalline solar cells are cheaper but less efficient. The process of manufacturing silicon with a single crystal structure makes it comparatively more expensive than polycrystalline silicon. The thin-film solar cell would be ideal to replace the Si cell, but it has suffered from efficiency problems. Thus, today, PV laboratories are devoting great effort to finding a substitute for silicon solar cells with good efficiency and lower cost and also trying to enhance the efficiency of current Si solar cells. A suitable approach has now been identified by researchers to improve cell efficiency by using an upconversion/downconversion (UC/DC) phosphor layer coating on the front substrate of the Si cell, which will be discussed in the next chapter.

References

1. Atom. *Compendium of Chemical Terminology (IUPAC Gold Book)*, (2nd edn). IUPAC. Retrieved 25 April (2015).
2. D. C. Ghosh, R. Biswas, Theoretical calculation of absolute radii of atoms and ions. Part 1. The atomic radii. *Int. J. Mol. Sci.* 3 (2002) 87–113.
3. G. J. Leigh, *International Union of Pure and Applied Chemistry, Commission on the Nomenclature of Inorganic Chemistry, Nomenclature of Organic Chemistry – Recommendations*. Oxford, UK: Blackwell Scientific Publications. ISBN 0-08-022369-9) (1990) 35.
4. A. W. Thackray, The origin of Dalton's chemical atomic theory: Daltonian doubts resolved. *Isis*. The University of Chicago Press on behalf of The History of Science Society (ISSN 0021-1753) 57 (1966) 35.
5. T. Thomson, *The Elements of Chemistry*, J. & A. Y. Humphreys. Philadelphia: Published and Sold (1810) 480.

6. H. E. Roscoe, H. Arthur, *A New View of the Origin of Dalton's Atomic Theory*. London: Macmillan. (ISBN 1-4369-2630-0) (Retrieved 24 December 2007). (1896).

7. T. Levere, *Transforming Matter: A History of Chemistry from Alchemy to the Buckyball*. Baltimore, Maryland: The Johns Hopkins University Press. (ISBN 0-8018-6610-3) (2001) 84.

8. L. Akhlesh, E. E. Salpeter, Models and Modelers of Hydrogen. *Am. J. Phys.* World Scientific. (ISBN 981-02-2302-1) 65 (9) (1996) 933.

9. E. Rutherford, The scattering of α and β particles by matter and the structure of the atom, *Philos. Mag.*, 21 (6) (1911) 669.

10. G. Nicholas, *College Physics: Reasoning and Relationships*. Cengage Learning. (ISBN 1-285-22534-1.) (2012) 1051.

11. M. A. B. Whitaker, The Bohr–Moseley synthesis and a simple model for atomic x-ray energies. *Eur. J. Phys.* 20 (1999) 213.

12. Z. Constan, Learning nuclear science with marbles. *Phys. Teacher* 48 (2) (2010) 114.

13. B. Niels, On the constitution of atoms and molecules, part II systems containing only a single nucleus (PDF). *Philos. Mag.* 26 (153) (1913) 476.

14. C. Kittel, *Introduction to Solid State Physics*, 7th edn. Wiley, New Dehli, India (ISBN 0-471-11181-3) (1995).

15. W. Shockley, *Electrons and Holes in Semiconductors: With Applications to Transistor Electronics*. R. E. Krieger (ISBN 0-88275-382-7) (1950).

16. N. Donald, Semiconductor physics and devices (PDF). Elizabeth A. Jones.

17. F. Feynman, *Feynman Lectures on Physics*. Basic Books (1963).

18. A. Al-Azzawi, *Light and Optics: Principles and Practices*.CRC Press, Boca Raton, FL. 2007. 4 March (2016).

19. K. V. Singh, K. Jerath, Thermoelectric Cooler, *International Journal of Mechanical and Industrial Technology*, vol. 4 (2016) 78.

20. D. Attwood, B. Hartline, R. Johnson, The Advanced Light Source: Scientific Opportunities, Lawrence Berkeley Laboratory Publication 55 (1984).

21. F. C. Brown, *Solid State Physics*, vol. 29, H. Ehrenreich, F. Seitz, D. Turnbull (eds.), Academic Press, New York, NY, (1974) 1.

22. B. J. Orr, J. G. Haub, Y. He, R. T. White. *Spectroscopic Applications of Pulsed Tunable Optical Parametric Oscillators*. In Duarte F. J. Tunable Laser Applications (3rd ed.). Boca Raton: CRC Press. (ISBN 978-1-4822-6106-6) (2016) 17–142.

23. A. V. Nurmikko, *Semiconductors and Semimetals*, vol. 36, D. G. Seiler, C. L. Littler (eds.), Academic Press, New York, NY, (1992) 85.

24. R. R. Alfano (ed.), *Semiconductors Probed by Ultrafast Laser Spectroscopy*, vols. I and II, Academic Press, New York, NY, (1984).

25. G. Bastard, C. Delalande, Y. Guldner, P. Vosin, in *Advances in Electronics and Electron Physics*, vol. 72, P. W. Hawkes (ed.), Academic Press, New York, NY, (1988) 1.

26. C. Weisbuch, B. Vinter, *Quantum Semiconductor Structures, Fundamentals and Applications*, Academic Press, New York, NY, (1991) 57.

27. M. Born, K. Huang, *Dynamical Theory of Crystal Lattices*, chapter 2, Oxford University Press, London, (1954) 38.

28. W. K. H. Panofsky, M. Phillips, *Classical Electricity and Magnetism*, Addison-Wesley, New York, NY, (1962) 29.

29. J. M. Ziman, *Principles of the Theory of Solids*, Cambridge University Press, London, (1972) 200.

30. D. L. Greenaway, G. Harbeke, *Optical Properties and Band Structure of Semiconductors*, Pergamon, London, (1968) 9.

31. C. Kittel, *Introduction to Solid State Physics*, 4th edn, Wiley, New York, NY, (1971) 184.

32. S. S. Mitra, N. E. Massa, *Handbook on Semiconductors*, T. S. Moss, W. Paul (eds.), North Holland, Amsterdam, (1982) 81.

33. W. G. Spitzer, *Semiconductors and Semimetals*, vol. 3, R. K. Willardson, A. C. Beer (eds.), Academic Press, New York, NY, (1967) 17.

34. J. L. Birman, *Theory of Crystal Space Groups and Lattice Dynamics*, Springer-Verlag, Berlin, (1974) 271.

35. E. A. Irene, H. Z. Massoud, E. Tierney, Silicon oxidation studies: Silicon orientation effects on thermal oxidation, *J. Electrochem. Soc.*, 133 (1986) 1253.

36. G. Eranna, *Crystal Growth and Evaluation of Silicon for VLSI and ULSI*. CRC Press. (ISBN978-1-4822-3282-0) (2014) 253.

37. J. Guldberg, *Neutron-Transmutation-Doped Silicon*. Springer Science & Business Media. (ISBN978-1-4613-3261-9) (2013) 437.

38. C. M. Parry, Bismuth-doped silicon: An extrinsic detector for long-wavelength infrared (LWIR) applications. *Mosaic Focal Plane Methodologies I* 0244 (1981) 2.

39. H. S. Rauschenbach, *Solar Cell Array Design Handbook: The Principles and Technology of Photovoltaic Energy Conversion*. Springer Science & Business Media. (ISBN978-94-01107915-7) (2012) 157.

40. I. Weinberg, H. W. Brandhorst Jr. Lithium counterdoped silicon solar cell. U.S. Patent 4,608,452 (1984).

41. M. Ohring, *Reliability and Failure of Electronic Materials and Devices* (ISBN 0-12-524985-3) Academic Press (1998) 310.

42. S. A. Ahmed, Prospects for photovoltaic conversion of solar energy. In *Alternative Energy Sources*, J. T. Manassah (ed.) Elsevier (1980) 365.

43. U.S. Geological Survey (2008). Germanium—statistics and information. U.S. Geological Survey, Mineral Commodity Summaries. Retrieved 28 August (2008).

44. B. Sheila, G. R. Ryne, E. Keith, Space and terrestrial photovoltaics: Synergy and diversity. *Progr. Photovoltaics Res. Appl.* 10 (6) (2002) 399.

45. Germanium. Los Alamos National Laboratory. Retrieved, 28 August (2008).

46. Solar Insight, Research note—PV production (2013): An all Asian-affair (PDF). Bloomberg New Energy Finance. 16 April (2014) 2, Archived from the original on 30 April (2015).

47. China: The new silicon valley—Polysilicon. 2 February 2015. Archived from the original on 30 April (2015).

48. Fraunhofer ISE. Photovoltaics Report, July 28, (2014) 24.

49. Fraunhofer ISE. Photovoltaics Report, July 28, (2014) 23.

50. Solar Energy Perspectives: Executive Summary. International Energy Agency. 2011. Archived from the original (PDF) on 3 December 2011. X. P. Chen, X. Y. Huang, Q. Y. Zhang, *J. Appl. Phys.* 106 (2009) 063518.

51. G. Peter, *Sustainable Energy Systems Engineering: The Complete Green Building Design Resource*. McGraw Hill Professional. (ISBN 978-0–07-47359-0) (2007).

52. B. van der Zwaan, A. Rabl, Prospects for PV: a learning curve analysis. *Sol. Energy* 74 (2003) 19.

53. O. Morton, Solar energy: a new day dawning? Silicon Valley sunrise. *Nature* 443 (2006) 19.

54. A. Goetzberger, C. Hebling, H. W. Schock, Photovoltaic materials, history, status and outlook. *Mater. Sci. Eng. R.* 40 (2003) 1.
55. W. Shockley, H. Queisser, Detailed Balance Limit of Efficiency of p-n Junction Solar Cells. *J. Appl. Phys.* 32 (1961) 510.
56. B. Richards, Luminescent Layers for Enhanced Silicon Solar Cell Performance: Down-Conversion. *Sol. Energy Mater. Sol. Cells* 90 (2006) 1189.
57. B. van der Zwaan, A. Rabl, Prospects for PV: a learning curve analysis. *Sol. Energy* 74 (2003) 19.
58. W. R. Taube, A. Kumar, R. Saravanan, P. B. Agarwal, P. Kothari, B. C. Joshi, D. Kumar, Efficiency enhancement of silicon solar cells with silicon nanocrystals embedded in PECVD silicon nitride matrix. *Sol. Energy Mater. Sol. Cells* 101 (2012) 32.
59. L. Q. Zhu, J. Gong, J. Huang, P. She, M. L. Zeng, L. Li, M. Z. Dai, Q. Wan, Improving the efficiency of crystalline silicon solar cells by an intersected selective laser doping. *Sol. Energy Mater. Sol. Cells* 95 (2011) 3347.
60. J. Zhao, A. Wang, M. A. Green, F. Ferrazza, 19.8% efficient "honeycomb" textured multicrystalline and 24.4% monocrystalline silicon solar cells. *Appl. Phys. Lett.* 73 (1998) 1991.
61. S. Kim, H. Lee, J.-W. Chung, S.-W. Ahn, H.-M. Lee, n-type microcrystalline silicon oxide layer and its application to high-performance back reflectors in thin-film silicon solar cells. *Curr. Appl. Phys.* 13 (2013) 743.
62. Y. H. Heo, D. J. You, H. Lee, S. Lee, H.-M. Lee ZnO:B back reflector with high haze and lowa bsorption enhanced triple-junction thin film Si solar modules. *Sol. Energy Mater. Sol. Cells* 122 (2014) 107.
63. S. Kim, J.-W. Chung, H. Lee, J. Park, Y. Heo, H.-M. Lee, Remarkable progress in thin-film silicon solar cells using high-efficiency triple-junction technology. *Sol. Energy Mater. Sol. Cells* 119 (2013) 26.
64. M. A. Green, *Third Generation Photovoltaics.* Springer-Verlag (ISBN 3-540-26562-7) (2003).
65. www.ni.com/white-paper/7229/en/, accessed 13 May 2018.

Part II

Phosphors—An Overview

2

Introduction to Phosphors, Rare Earths, Properties and Applications

2.1 Phosphors

It is well known that over 80 years ago, the concept of blackbody radiation came into existence and it would be used to illuminate our environment. After continuous efforts by scientists and researchers, lamps, TV sets, monitors, and medical scanners were developed and are now used everywhere. Solid inorganic luminescent materials play an important role in the development of this display and lighting technology. Currently, research in this area has been conducted for almost a hundred years, but the search for new phosphor materials continues. Phosphors are defined as solid luminescent materials that have the ability to absorb incident light (in the near ultraviolet [UV] or visible region) and then to emit light in a longer wavelength range (visible to near infrared [NIR]). Generally, a xenon flash lamp is used as the excitation source (i.e., the process of photoluminescence), or it may be an incident beam of electrons (a process called *cathodoluminescence*). The phosphor is composed of an inert divalent or trivalent host lattice doped with a rare earth-transition series element. The activator impurity ions typically have a 3d or 4f electron in their outermost shell, which is then excited under a UV or visible light source. In the luminescence process, the activator absorbs the incident energy and goes to a higher state, where relaxation occurs. Subsequently, it emits a photon and returns to the lower state. The emission originates from the phosphor when it is doped with divalent or trivalent impurity ions, so that it forms different energy levels and shows UV and visible emission. The phosphor efficiency depends on the amount of light absorbed and the amount of light emitted, and for most of the ideal phosphors, this ratio is considered to be unity. Other properties associated with ideal phosphors are thermal stability, color quality, *color rendering index* (CRI), and so on. During the phosphor relaxation process, part of the energy is lost due to nanoradiative relaxation of photons. This is the important factor that affects the luminous efficiency of the phosphor. So, for better luminous properties, there is a need for reduced nonradiative loss. At present, there are very few phosphor materials used on a commercial level for the fabrication of white light–emitting diodes (LEDs), compact fluorescent lights (CFLs), and display devices. Blue-, green-, yellow-, and red-emitting phosphors are the most used primary phosphors in many display devices. With a preferred combination of these, we can obtain a white emission. Thus, the application of phosphors may change according to their photoluminescence and related properties, which are found to be close to the ideal phosphor. In the third generation of photovoltaic (PV) technology, NIR-emitting phosphors were experimentally and practically proposed by many researchers to enhance the light conversion efficiency of solar cell devices. Hence, phosphor research is a leading field for researchers, having better technological advantages, such as energy-saving and energy-conversion capacity. In the next section, we will discuss some important terms that are related to luminescence, phosphors and their mechanisms, rare earth–transition spectroscopy, and so on.

2.1.1 Luminescence

Basically, light can be categorized in one way as incandescence and in another way as luminescence. Incandescence is a phenomenon in which solid materials emit light due to heating [1]. Therefore, luminescence is defined as the emission of light by bodies that is in excess of that attributable to blackbody radiation and persists considerably longer than the periods of electromagnetic radiations in the visible range after the excitation stops.

Luminescence occurs when atoms of the solid are energized without sufficient heating of the bulk material. Then, energized atoms liberate their surplus energy through either ultraviolet, visible or infrared radiation (Figure 2.1) [2]. Luminescence phenomenon have fascinated mankind for centuries. Light from the aurora borealis, glow worms, luminescent wood, and rotting fish and meat are all examples of naturally occurring luminescence [3]. The word *luminescence* is derived from the Latin word *lumen*, which means "light." In

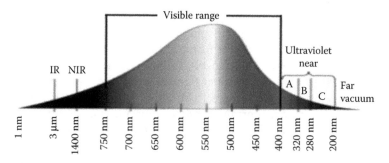

FIGURE 2.1
Visible light spectral region.

1888, a German physicist, Eilhardt Wiedemann, used the word *luminescence* for the first time. Luminescence is a collective term generally used for substances emitting light in the absence of heat, which is also reflected by the term *cold light*. After this fundamental discovery, similar and equivalent inventions were reported from all over the world. Materials showing luminescence are known as *luminescent materials* or *phosphors* (meaning 'light bearer' in Greek and appearing as the symbol of the morning star Venus in the Greek tradition). The term *phosphor* was first used in the seventeenth century by an Italian alchemist named Vincentinus Casciarolo from Bologna [4, 5]. The different facts of luminescence depend on the nature, concentration, spatial distribution, mutual interactions of the types of lattice imperfection.

2.1.2 Mechanism of Luminescence

The luminescent materials called phosphors are basically inorganic solid white-colored materials. They are composed of a host lattice and a luminescent center, frequently called an *activator*.

The basic mechanism of the luminescence excitation and emission process is depicted in Figure 2.2, in which A and A* indicate the ground and excited state of an atom, respectively. The blue arrow indicates the excitation of the atom to the higher energy level after absorbing the incident energy, while R and NR show the radiative and nonradiative paths of the atom from the excited state toward the ground state. Therefore, when this atom returns rapidly to the ground state, it emits radiation of a certain wavelength that is longer than the excitation wavelength. In some luminescence hosts, the excitation wavelength is not absorbed by the activator ions unless an impurity ion has been added that can absorb the excitation wavelength and afterward transfer it to the activator ion. In this case, the absorbing ion is called a *sensitizer*. In several cases, the host lattice itself acts as the sensitizer by transferring its excitation energy to the activator. The mechanism of luminescence

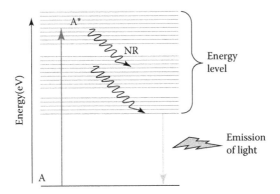

FIGURE 2.2
Basic mechanism of luminescence.

can be explained with the help of a configurational coordinate diagram as shown in Figure 2.3 [6]. A small number or a combination of specific normal coordinates can approximate the huge number of actual vibrational modes of the lattice. These normal coordinates are called the *configurational coordinates*. Here, the total energy (the sum of the electron energy and ion energy) of the activator ion is plotted against the distance between metal cations and anions in the host lattice denoted by 'r'. 'r$_0$' stands for the equilibrium distance of the ground state activator and 'r$_1$' for the equilibrium distance of the excited state activator.

The excitation (λ_{ex}) and emission (λ_{em}) transitions are shown by vertical blue and red arrows, which reveals that the nucleus of the activator remains at the same position throughout the optical process. This is called the *Franck–Condon* principle, which shows that an atomic nucleus is 10^3 to 10^5 times heavier than an electron and the electron moves much faster than the nucleus. At r$_0$, the optical absorption proceeds from the equilibrium position of the ground state. The probability for an excited electron to lose energy by generating lattice vibration is 10^{12} to 10^{13} s^{-1}, while the probability for light emission is at most 10^9 s^{-1}. As a result, State B relaxes to the equilibrium position C before it emits luminescence. This is followed by the emission and relaxation process D→A, completing the cycle. When the system goes from D to A, heat energy is given up, and the transition is nonradiative. This is the cause of Stokes shift observed in the host lattice. When the system is at an equilibrium position at Point C, the upper excited curve is not at rest but migrates over a small region around C due to the thermal energy of the system. With a further increase in temperature, these fluctuations cover a wider range of the configuration coordinate. As a result, emission occurs via downward transition to Point D, but it is not restricted to Point D on the ground

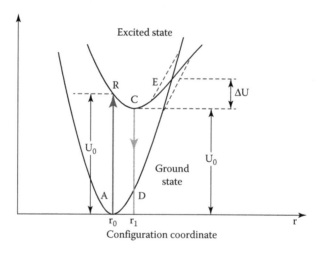

FIGURE 2.3
Configurational coordinate diagram.

state curve; it also covers a region around D. The ground state curve shows a rapid change of energy; therefore, for small values of the configuration coordinates, it leads to a large range of energies in the optical transition, so that it gives a broad absorption and emission band. From this, it is observed that the widths of the band should vary with respect to temperature, and this relationship is valid for temperature above and below room temperature. The phenomenon of luminescence quenching and the variation of decay time of luminescence due to temperature can also be studied with the help of this model. Further, due to the occurrence of intense thermal vibrations, the system may be raised to Point E, which is the point of intersection of the excited vibrational energy and ground state, from which the atom can return to the ground state by emitting a small amount of heat or infrared radiation. For centers that stay in the excited state for a relatively longer time, the temperature quenching tends to occur most strongly. As a result, the decay time of the emission that occurs in this temperature region is largely characteristic of centers in which transitions to the ground state take place quickly, and hence, the decay time of the luminescence decreases.

If Point E is at energy E_Q above the minimum of the excited state, the luminescence efficiency η is given by

$$\eta = \left[1 + C \exp\left(-E_q/kT \right) \right]^{-1} \tag{2.1}$$

where:

C is a constant

k is the Boltzmann's constant

T is the temperature in Kelvin

$$\text{Therefore for fluorescence emission } 1/t = P_r + P_{nr} \tag{2.2}$$

where:

t is the mean lifetime

P_r is the probability of radiative transition, with little dependence on temperature

P_{nr} is the probability of nonradiative transitions increasing with temperature

If there are no nonradiative transitions at zero temperature, the lifetime is given by

$$1/t_0 = P_r \tag{2.3}$$

$$\eta(T) = t(T)/t(0) = P_r / (P_r + P_{nr}) \tag{2.4}$$

Also, if the equilibrium position of the excited state C is located outside the configuration coordinate curve of the ground state, the excited state intersects with the ground state in relaxing from B to C, which leads to a nonradiative process.

2.1.3 Classification of Luminescence

Depending on the duration of persistence, luminescence can be further classified into fluorescence and phosphorescence. Fluorescence is characterized by temperature-independent decay. This process has a relatively short lifetime to emit light (10^{-6} to 10^{-12} s). Also, it corresponds to the emission of visible light by a material under the stimulus of visible or invisible radiation of shorter wavelength. Phosphorescence is characterized by temperature-dependent decay, which persists considerably longer (even for seconds). If the fluorescence persists for an appreciable time after the stimulus has been removed, this afterglow is called *phosphorescence*, which sometimes may differ in color from the original fluorescence [7]. Luminescence can be further classified on the basis of the excitation source. Table 2.1 shows the types of luminescence and their application in lighting, medical, display devices, and PV technology.

2.2 Rare Earth Ions

Rare earth elements are a set of 17 chemical elements in the periodic table. Specifically, the 15 lanthanides ions are considered as rare earth elements,

TABLE 2.1

Classification of Luminescence and Its Various Applications

Type of Luminescence	Excitation Source	Advantages
Photoluminescence	Photons	• Fluorescent lamps • PL-LCD • Upconversion • Downconversion
Electroluminescence	Electric potential	• LEDs • EL displays • Diode lasers
Thermoluminescence	Electric heating	• Radiation dosimetry • Archeological • Geological dating
Chemoluminescence	Chemical reaction	Analytical chemistry
Cathodoluminescence	Electron impact	• FED • Oscilloscope
Bioluminescence	Biochemical reaction	Analytical chemistry
Radioluminescence	Bombardment of ionizing radiations	• X-ray imaging • Scintillators

plus scandium and yttrium, since they exhibit similar chemical properties. They are found in either +2 or +3 oxidation state. In 1788, Geijer reported a black stone found close to the Swedish town of Ytterby [8]. Hence, the stone was called *Yttria*. Then, in 1803, Klaproth reported a stone named *Ceria*. It was found that both materials were derived from gadolin. It took about a decade to analyze them, as both materials were found to include a number of different but chemically very similar elements, which made it hard to isolate them. Lanthanum, cerium, samarium, europium, and gadolinium were derived from Ceria [9]. Praseodymium and Neodymium were derived from Didymium, while terbium, erbium, ytterbium, and holmium were derived from Yttria. Rare earth ions are also called *lanthanide ions*, which play an important role in modern display technology as well as acting as active constituents in many optical materials. These materials have gained more interest and applications in many solid-state lighting and display devices; this is a growing technology of the twenty-first century due to its global advantages and energy-saving capacity. Hence, rare earth ions are an optically active element in many types of solid-state luminescent materials. The optical properties of materials are influenced by the interactions between the electronic band states of the host lattice and the rare earth ions. Among the different lanthanide elements (from La to Lu, about 15 elements), still only a few rare earth ions have been used in display materials in the past few years. These ions have numerous energy levels, and the emission arises from their forbidden or allowed transitions. Rare earth ions having transitions such as the $4f^N$ to $4f^N$ spectroscopic transition show long lifetimes, sharp absorption lines, and excellent coherence properties, while other ions that show $4f^N$ to $4f^{N-1}$ 5d transitions have large oscillator strengths, broad absorption bands (from the ultraviolet to the visible region), and short lifetimes. The broadband-emitting $4f^{N-1}$ 5d states of the rare earth ions have many applications in display and solid state lighting (SSL) technology. The relative energies of the host lattice may be helpful to determine the binding energies of the $4f^N$ states relative to the host bands, and they are also used to measure the energy differences between $4f^N$ and $4f^{N-1}$ 5d states [2]. The barycenters of $4f^N$ to $4f^{N-1}$ 5d transition energies show large variations from material to material. They may be determined from the absorption spectra of each rare earth ion when it is doped with host material. Also, it is well known that the binding energies can be expressed as the energy required to remove the 5d electron from the $4f^{N-1}$ 5d configuration, leaving the tetravalent ion with a $4f^{N-1}$ configuration in the lowest energy state. The lanthanides from Ce^{3+} to Lu^{3+} have from 1 to 14 electrons in the 4f state, which is equivalent to [Xe]. A series of elements start from La^{3+} and Lu^{3+} (as shown in Table 2.2) have 0 to 14 electrons in its shell, and hence, no electronic energy levels can induce excitation and luminescence processes in the visible or NIR region. From the lanthanide ions' spectroscopic properties, most rare earth elements are filled with one or more electrons in their outermost 4f orbital. Thus, different rare earth (divalent/trivalent ions) elements have abundant energy levels

TABLE 2.2

Different Rare Earth Elements and Their Symbols

Rare Earth Element	Symbol	Atomic No.
Scandium	Sc	21
Yttrium	Y	39
Lanthanum	La	57
Cerium	Ce	58
Praseodymium	Pr	59
Neodymium	Nd	60
Promethium	Pm	61
Samarium	Sm	62
Europium	Eu	63
Gadolinium	Gd	64
Terbium	Tb	65
Dysprosium	Dy	66
Holmium	Ho	67
Erbium	Er	68
Thulium	Tm	69
Ytterbium	Yb	70
Lutetium	Lu	71

due to differences in their arrangement of 4f electrons. When 4f outermost electrons transit among different energy levels, they can produce numerous absorption and fluorescence spectra. Therefore, most rare earth elements are often used as dopants in many light-emitting materials and laser materials to study their emission properties in the ultraviolet and visible regions of the spectrum [3]. The general properties of the rare earths have been well discussed on the basis of allowed or forbidden electronic transitions; these are well known, and their theoretical description is also well discussed in the literature [4]; therefore, the study of the location of energy levels and their energies of transition has attracted great interest in the past few years. The electronic structures of rare earth activated insulators, which incorporate the relationships between the location of energy levels and their states, are important to understand the interactions between these rare earth energy states and their influence on the properties of materials. Nowadays, rare earth–based oxides are available as high-purity starting materials, but it is essential to identify which impurity ions can give better luminescence efficiency. Very few rare earth ions show the broad excitation bands that correspond to transitions with high absorption strength belonging to the fluorescent center itself. Two types of transitions can be distinguished. The first is known as a *charge-transfer transition* from ligands (usually 02- ions) to the central rare earth ion, and the second is interatomic 4f–5d transitions. These will be discussed in the next section with the help of excitation and emission spectra of rare earth-activated phosphor under NUV excitation wavelength.

The rare earth ions are characterized on the basis of the innermost unfilled 4f shell. This 4f shell lies inside the ion and is shielded well from the surrounding filled $5s^2$ and $5p^6$ orbitals. Therefore, there is little influence of the host lattice on the optical transitions within the $4f^n$ configuration. Figure 2.4 shows the Dieke energy level diagram originating from the $4f^n$ configuration as a function of n for the trivalent ions. The width of the bars indicates order of magnitude of the crystal field splitting [10]. This diagram shows the energy levels for trivalent rare earth ions. It is important to know the formation of energy levels when we dope rare earth ions in a new host lattice. Also, it is an essential tool in the design of materials suitable for energy-saving and energy-conversion purposes. Optical transition of divalent and trivalent rare earth ions is parity allowed (e.g., Eu^{2+}, Ce^{3+}). The f–d transition has five or six components that depend on crystal field splitting of the host lattice as shown in Figure 2.5. From the configurational coordinate model,

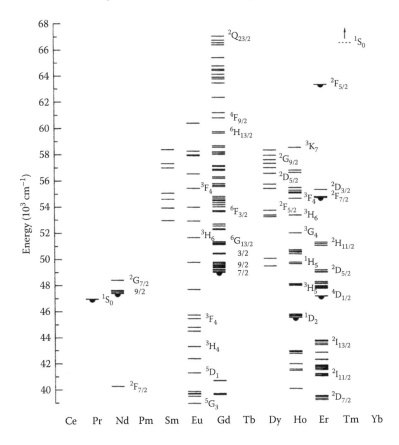

FIGURE 2.4
Dieke diagram indicating the energy levels in rare earth ions. (From R.T., Wegh, A. Meijerink!, R. J. Lamminmäki, and J. Hölsä. Extending Dieke's diagram. *J. Lumin.*, 87–89, 1002, 2000, Copyright (2000). With permission from **Elsevier.**) [10]

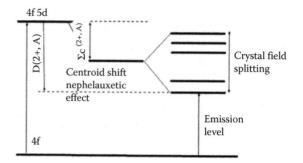

FIGURE 2.5
Energy level in 4f–5d transition of rare earth ions.

both have $R \neq 0$ and the spectra observed in the broad absorption bands. The allowed and forbidden transitions in rare earth ions will be discussed in detail in the next section.

2.3 Transitions in Rare Earth Ions

2.3.1 Allowed Transitions

Optical absorption spectra of rare earth ions with $4f^N$ configuration are often due to 4f-to-5d (f–d) transitions, that is, the optical transfer of an electron from the 4f to 5d shells of the ions (i.e. from $4f^n - 4f^{n-1}5d$), or charge-transfer (CT) transitions, that is, promotion from the ligand atom to the unfilled 4f shell ($4f^n-4f^{n+1}$). In comparison with the f–f transition, the f–d transitions are not understood well experimentally or theoretically from the observed spectra, because the f–d transitions of the trivalent rare earth ions generally lie in the UV or vacuum ultraviolet (VUV) region, making them experimentally less accessible. The phosphors used for the fabrication of fluorescent lamps [11], plasma display screens, and mercury-free light tubes have high quantum efficiency, up to 90% [12–14]. There are many reasons why these promising cascading routes have not been successfully used in developing lamp phosphors. One of them is the unpredictability of the position of the $4f^n-4f^{n-1}5d^1$ band of the rare earth ions. Unlike the forbidden transitions with the f-manifold of the rare earth ions, this transition is parity allowed. The position of the emission band depends on the host lattice due to the strong covalency of 5d-like levels. The $4f^{n-1}$ 5d transitions are characterized by the strong broadband. Frank–Condon factors and their absorption and emission spectra are broadband due to the high density of states of the $4f^{n-1}$ 5d electronic configuration. The f–d transitions usually occur in broad spectral bands with very little structure, even though there are a large number of energy levels involved in the f–d transitions. The structures of the f–d transition spectra are quite different for a particular ion in different host lattices due to crystal field splitting, but in many cases, there are similarities for different ions in the same host. There have been some measurements on f–d transitions for Eu^{2+}, Sm^{2+}, and Yb^{2+} in halide crystals and for most trivalent ions in $LiYF_4$, CaF_2, and YPO_4 hosts. The f–d absorption spectra of Eu^{2+} have been studied by many authors [15]. The spectra contain two broad bands around 27,000 and 43,000 cm^{-1}. The low energy band has a staircase structure of about seven peaks, with intensity increasing from low to high energy. Eu^{2+} in a CaF_2 host contains two broad bands in the UV region, with a "staircase" structure for the band with lower energy. The two broad bands were assigned to a transition [16] from the $4f^7$ ground state to the $4f^6 5d^1$ excited states with the d electron in e_g and t_{2g} orbitals. The staircase was assigned to the splitting of the ground states of $4f^6$ core by spin-orbit interaction.

Among the different activators in 5d–4f transition spectroscopy, Ce^{3+} and Eu^{2+} ions possess f–d electron configurations, and they are good candidates as activators in phosphors because they can emit broadband visible light due to the influence of crystal field and nephelauxetic effects, as shown in Figure 2.5 [17–19]. Additionally, due to the difference between the intrinsic electronic configurations of Ce^{3+} and Eu^{2+} ions, the energy of the d→f transition of the former is higher than that of the latter when they enter the same lattice site in a particular phase; thus, Ce^{3+} can transfer its absorbed energy to Eu^{2+} [20, 21].

2.3.2 Forbidden Transitions

The transition from $^5D_0 \rightarrow ^7F_0$ is spin and electric dipole forbidden. J is zero for the ground state $^7F_0$ and also for the principal emitting state $^5D_0$. Accordingly, these states cannot show stark splitting. In Eu^{3+}, the spin selection rule is relaxed by the mixing of $^7F_0$ into $^5D_0$ by the spin-orbit interaction. The $^5D_0 \rightarrow ^7F_0$ transition is nondegenerate and not subject to crystal field splitting (CFS) changes in the vicinity of the Eu^{3+} ion. Therefore, the position and line width of this 0↔0 transition give information about the local bonding environment and the coordination of the Eu^{3+} ion in the host lattice [22]. The $^5D_0 \rightarrow ^7F_0$ transition is ideally suited for investigating line broadening in glasses because of a single transition between nondegenerate energy levels. Also, it can be studied with no overlapping from neighboring crystal field components. The $^5D_0 \rightarrow ^7F_0$ transition is taken as a standard for the inhomogeneous nature of glass, since both $^5D_0$ and $^7F_0$ states are nondegenerate [23]. Thus, factors such as local crystal fields, electron–phonon coupling, and ion–ion interaction may contribute to this type of broadening. Moreover, the energy gap between the $^5D_0$ excited state and the next $^7F_6$ lower state is about ten times higher than the phonon energy involved. Therefore, the probability of nonradiative deexcitation of the $^5D_0$ excited state by the multiphonon process is very small [24]. The $^5D_0 \rightarrow ^7F_0$ line is asymmetric when the J mixing mechanism is dominant and nearly symmetric when other mechanisms are dominant. The inhomogeneous broadening of the $^5D_0 \rightarrow ^7F_0$ transition is mainly determined by the energy fluctuation of the $^7F_0$ state among sites. This is because the crystal field potential causes mixing of the $^7F_2$ state with $^7F_0$, bringing about the lowering of the $^7F_0$ state, while for the $^5D_0$ state, the admixture of other states is much less than expected for the $^7F_0$ state. When the inhomogeneous broadening is large, the energy transfer between the same electronic states encounters a large energy mismatch. Thus, inhomogeneous broadening enhances the emission peak widths of the $^5D_0$ state, and the broadening of the $^5D_0 \rightarrow ^7F_0$ line is mainly due to crystal field level induced by J mixing. The Eu^{3+} ion takes a wide variety of coordination numbers, from 8 to 12 in oxide crystals and also in glasses [25]. Therefore, inhomogeneous broadening is dominant in this transition, and the absorption and emission line shapes coincide with each other. The shape of the line spectra reflects the statistical distribution of the transition energy.

2.4 Selection Rules

2.4.1 Laporte Selection Rule

The orbital quantum number (l) for electric dipole transitions changes according to $\Delta l = \pm 1$; thus, the initial and final states must have opposite parity. Due to several interactions with the electronic wave function of the ion, the Laporte selection rule can be relaxed for the following reasons.

1. *Electron–phonon interaction*

    In octahedral surroundings, even- and odd-type vibrational modes exist. Hence, even-type transitions secure a partly odd character and become allowed by coupling to vibrational modes.

2. *Interaction with higher orbitals*

    By mixing the wave function of higher orbitals (e.g., d and p orbitals), the odd character can be added to the wave function. This interaction depends on the symmetry, and it is commonly known for noncentrosymmetric surroundings. In local surroundings without inversion symmetry, the Laporte selection rule is further relaxed by the admixture of states having opposite parity compared with the $4f^n$ states. As a result, f–f transitions are observed to be much weaker on lattice sites having inversion symmetry.

2.4.2 Spin Selection Rule

During the transition between energy levels, the electron cannot change its spin; that is, spin is to be conserved ($\Delta S = 0$). Therefore, the wave function of the electron cannot be factorized strictly to $\psi = \psi_{space} \cdot \psi_{spin}$. Thus, spin-forbidden transitions are very weak for the spectra of transition metal complexes, but the intensity of the spectral line is increased due to strong spin-orbit coupling, and the spin selection rule is relaxed due to the presence of spin-orbit coupling.

2.4.3 Selection Rule for the Total Quantum Number J

The term J is called *total orbital angular momentum*, which can represented by the spectroscopic notation as ^7F_J. For the outermost 4f orbital of the rare earth ions, the selection rule is $|\Delta J| \leq 6$, which is derived by Judd and Ofelt theory [26, 27] from the calculations of the intensities of the spectral lines. For electric dipole transitions, having J→J transitions is strictly forbidden (i.e., J=0→0). The transitions J=0→1, 3, 5 appear extraordinarily weak in rare earth ions with an even number of electrons [28]. The parity selection cannot be relaxed if the rare earth ion is located at the center of symmetry in the relevant crystal lattice. For the magnetic diploe transition the selection rule assigned is $\Delta J = 0, \pm 1$ (except that J=0→0 is forbidden). Magnetic dipole f-f transitions are not affected much by the site symmetry because of their parity-allowed nature. For the allowed electric dipole transitions, oscillator strength is of the order of 10^{-5} to 10^{-6}, while for magnetic dipole transitions, it is 10^{-8}.

The lanthanide series consists of 14 elements, called *lanthanons* or rare earth elements. These are characterized on the basis of their outermost 4f energy level. From the electronic structure and oxidation states of the lanthanides, it is seen that most of the elements are more stable in the 3+ oxidation state. Lanthanum has the electronic structure: [Xe] $5d^1 6s^2$. Thus, it is expected that the 14 elements of these series, from cerium to lutetium, would be formed by adding 1, 2, 3, ... 14 electrons into the 4f level.

Rare earth ions without 4f electrons, that is, Sc^{3+}, Y^{3+}, La^{3+}, and Lu^{3+}, have no energy levels from where the rare earth ions shows absorption and emission process in near UV and visible region of the spectra. Other rare earth ions, from Ce^{3+} to Yb^{3+}, contain a partially filled 4f orbital, form the characteristic energy levels when doped into the host lattice, and show different types of luminescence properties. Many of these trivalent ions are used as active rare earth ions in many types of phosphor materials due to their allowed and forbidden transitions in the visible region, and they will be preferred to trivalent host cations such as Y^{3+}, Gd^{3+}, La^{3+}, and Lu^{3+} in various phosphor compounds (e.g., YAl_3O_5). According to the concept of atomic spectroscopy, the azimuthal quantum number (*l*) of the 4f orbital is 3, so that it gives rise to 7 $(2l+1)$ orbitals, each of which can accommodate at least two electrons. In the ground state of the ions, electrons are distributed to give maximum combined spin angular momentum (S). If the spin angular momentum of the electron is added to the orbital momentum (L), it gives the total angular momentum (J) of the electron, and it can be represented as follows:

$J = L - S$, where the number of 4f electrons is lower than 7

$J = L + S$, where the number of 4f electrons is higher than 7

The electronic state of the electron is denoted by specific spectroscopic notation as $^{2s+1}L_j$, where L is denoted by the S, P, D, F, G, H, I, K, L, M ... levels, which correspond to the values L = 0, 1, 2, 3, 4, 5, 6, 7, 8, 9, ..., respectively.

Thus, the actual electronic state of the electron in rare earth ions is considered as an intermediate state. More accurately, it is a mixed state, that is, $^{2s+1}L_j$ states that are combined by spin-orbit interaction. For more qualitative discussion, here the principal state (L) of the electron can be taken as the actual state. The mixing of the states that is observed due to spin-orbit interaction is small for the levels that exist near the ground states. However, it is considerable for higher states, which are the neighboring states with similar J values. Thus, the effect of mixing on the energy levels is relatively small, but the effect on their optical transition probabilities can be large. The 4f levels of each Ln^{3+} ion are characteristic in nature. The environment of the host lattice does not affect the 4f energy levels of electrons, because they are well shielded by the external electronic state of the outermost $5s^2$, $5p^6$ electrons. The characteristic energy levels of the 4f electrons of trivalent lanthanide ions have been previously investigated and studied by Dieke and co-workers. The energy levels may be divided into three categories, corresponding to the $4f^n$ configuration (sharp line f–f transitions), the $4f^{n-1}5d$ configuration (broad energy bands), and those corresponding to charge transfer involving neighboring ions.

2.5 Nature of Bands

The applications of Ln^{3+} ions mostly depend on their observed characteristics: luminescence spectra, quantum efficiency, stability, and color quality when doped into a suitable host lattice. Thus, phosphors with characteristic sharp and broad bands are assigned to allowed and forbidden transitions in Ln^{3+} ions. This plays an important role in the selection of phosphors for lamps and other display devices.

2.5.1 Sharp Bands

Transitions within the 4f levels are strictly forbidden, because the parity does not change. In addition to this, there exists an spin exclusion; that is, $\Delta s = 0$ are allowed. To analyze rare earth luminescence on the basis of the magnetic dipole and electric quadrupole transitions, VanVleck et al. explained only a few of the observed emissions with this hypothesis. Emission corresponding to $\Delta j = 6$ was experimentally observed, which could not be explained.

Forbidden transitions are observed due to the fact that the interaction of rare earth ions with the crystal field or with the lattice vibrations can mix states of different parties into the 4f states. Although these mixed states make the transitions observable, their oscillator strengths remain relatively low (forced electric dipole transitions). f-f transitions of rare earth ions can be allowed when mixing of opposite parity configurations of the 4f5d state occurs due to the presence of odd parity crystal fields. In the second order, the electric dipole transition becomes parity allowed. The transitions corresponding to the even values of J ($0 \rightarrow 0$ excluded) increase in their oscillator strengths due to this effect. Transitions that are not allowed as an electric dipole may take place as a magnetic dipole. Therefore, magnetic dipole transitions obey the selection rules, as $\Delta L = 0$, $\Delta S = 0$, $\Delta l = 0$, and $\Delta J = 0$ ($0 \rightarrow 0$ excluded). The selection rule on ΔL and ΔS becomes weak due to spin-orbit coupling. Interaction of rare earth ions with lattice vibrations can mix the state of different parties into 4f states. Vibronic transitions of RE ions are observed due to the coupling of $4f^n$ state with a vibrational mode of the lattice, whereas in the first order, the coupling occurs only with IR vibrations to break the parity selection rule of the purely electronic f→f transitions.

2.5.2 Broad Bands

Rare earth ions form discrete and abundant energy levels when doped into the host lattice, and due to this, they show broad absorption and emission bands, as appeared in Eu^{2+} and Ce^{3+} ions due to their allowed transition. Generally, the absorption/emission bands in rare earth ions are classified into two groups. In the first group, one of the 4f electrons is raised to the higher 5d levels. Electron transitions from the configuration $4f^n$ to $4f^{n-1}5d$ are allowed, while in the second group of bands, there is the promotion of an electron from one of the surrounding ions to the 4f orbit of the central ion. This is called a *charge transfer state*, and it is denoted by $4f^n 2p^{-1}$. The $4f^{n-1}5d$ levels are formed by the electron in the 5d orbital, that is, e_g to t_{2g}, interacting with the $4f^{n-1}$ level. As a result of the strong crystal field effect on the 5d electron, the $4f^{n-1}5d$ configurations of rare earth ions in solids are very different from those of free ions. The $4f^n$ to $4f^{n-1}5d$ absorption in most of the divalent or trivalent ions shows strong bands corresponding to splitting into the crystal field components of the 5d orbital. Therefore, their spectra are similar when the ions are embedded in the same type of host, and the structure of 5d bands can be fitted to energy differences in the group multiplets of the $4f^{n-1}$ configuration.

2.5.3 Charge Transfer Bands

Charge transfer bands in trivalent rare earth ions such as Sm, Eu, Tm, and Yb were observed for the first time by Jorgenson et al. and were later studied for various rare earth ions. Jorgenson developed the theory to calculate the location of CT bands, which depends on the ligand. The energy decreases with the electronegativity of the ligand ion. In the case of Eu^{3+} ion, the CT band provides strong excitation. The CT transition is frequently described by an electron being transferred from one orbital to another. In contrast, this process may lead to a large spatial expansion of the charge distribution around the optically active centers rather than actual transfer of the electron. The CT state has a large width, which varies from 5,000 to 10,000 cm^{-1}, and it shows a large Stokes shift of several thousand wavenumbers. The band appearing in CT transition is very intense, and it follows the spin selection rule. The intensity of the bands depends on the displacement of the charge across the typical interatomic distance; thus, it produces a large transition dipole moment and oscillator strength. However, for intervalence CT transitions, only the Laporte selection rule is applicable. It has been observed that the energy of CT decreases with the electronegativity of the ligand ion. Tetravalent ions often show absorption in the visible region of the spectrum, which corresponds to the CT state. In the case of Eu^{3+} ion, the CT band provides strong excitation. No other rare earth ion has been investigated for the CT bands as much as Eu^{3+}.

2.6 Photoluminescence Excitation (PLE) and Emission Characteristics of Rare Earth Ions

Here, we discuss the photoluminescence excitation and emission characteristics of some rare earth ions that have been extensively used in many display devices for the past few years as an active center. A short description is provided for each ion, including the formation of energy levels and their spectroscopic transitions. Among the different elements in the lanthanide series, only a few RE ions show emission bands in the NIR range. They are widely used to study the up-/downconversion mechanism in a suitable host lattice due to their potential applications in crystalline silicon (c-Si) solar cells to improve the conversion efficiency.

2.6.1 Eu^{3+}

The Ln^{3+} ions are well characterized by a partially filled 4f shell that is shielded by the outermost $5s^2$ and $5p^6$ orbitals. The emission band appears in sharp and narrow lines, which are assigned to f–f transitions. Therefore, these rare earth ions are used on the basis of line type; it can be specific to the visible and NIR range, with high quantum efficiency and high lumen output [29]. Among the different trivalent ions, Eu^{3+} is the most frequently used activator ion in luminescent materials; it acts as a red component for the fabrication of many display devices. The Eu^{3+} state is more stable than Eu^{2+}. Eu^{3+} mainly shows emission bands in the red region due to transitions from $^5D_0 \rightarrow {}^7F_J$ (J = 1–4). Also, it is well known that Eu^{3+} ion is a good probe for the chemical environment of the lanthanide ion. The bands assigned to $^5D_0 \rightarrow {}^7F_1$ and $^5D_0 \rightarrow {}^7F_2$ correspond to the magnetic and electronic dipole transitions, respectively, and the relative intensities of the two emission bands depend strongly on the local symmetry of the Eu^{3+} ion. The nature of the electric and magnetic bands depends on the site symmetry. The magnetic dipole transition $^5D_0 \rightarrow {}^7F_1$ is dominating with inversion symmetry, while the electric dipole transition $^5D_0 \rightarrow {}^7F_2$ is stronger without inversion symmetry. However, from the energy level diagram of Eu^{3+} ions, it is seen that they have complex energy levels, which in turn depends on the host matrices. The first excited $^5D_0 \rightarrow {}^7Fj$ configuration is observed due to the large spatial extension of the 5D wave function. Other well-resolved bands are assigned to the $^5D_0 \rightarrow {}^7F_{0,1,2,3}$ transition of Eu^{3+} ions. The asymmetry factor, that is, I $(^5D_0 \rightarrow {}^7F_2)$ to I $(^5D_0 \rightarrow {}^7F_{0,1,3})$, is greater than 1, suggesting that Eu^{3+} ions are at a site lacking inversion symmetry. Sometimes, Eu^{3+} shows emission in the NIR range under UV excitation wavelengths, and this type of highly efficient sharp band may be suitable for PV applications as a prominent downconversion (DC) phosphor.

2.6.2 Sm^{3+}

Trivalent samarium (Sm^{3+}) is also an important activator ion used in the past few years in many display devices and white LEDs; also, it can show NIR DC properties. It exhibits a sharp orange-red photoluminescence spectrum corresponding to the visible and NIR region under UV excitation. Sm^{3+} ion shows a number of energy levels lying close to each other, which under UV excitation when doped with suitable host lattice renders the interpretation of the absorption spectra rather cumbersome. As discussed in the previous section, the outermost 4f electrons of Ln^{3+} ions are well shielded from the surroundings, and therefore, the emission is in sharp lines. However, their 4f states weakly interact with the host lattice, and thus, the energy differences are nearly constant, which leads to almost the same emission bands in different host lattices [30, 31]. The observed $^4G_{5/2}$ level in the visible region shows a relatively high quantum efficiencies (QEs) and different quenching channels. In general, Sm^{3+} can be easily reduced to Sm^{2+} under suitable experimental conditions. Sm^{3+} ions show several energy level intensities in many host lattices, and these are estimated quantitatively by the Judd–Ofelt (J–O) parameter, using the optical absorption and emission wavelengths of intra-configurational f–f transitions [32, 33]. The orange emission in Sm^{3+} ion consists of transitions from the $^4G_{5/2}$ excited state to the ^6H_J (J = 7/2, 9/2, and 11/2) ground state [34]. Thus, the transitions from $^4G_{5/2} \rightarrow {}^6H_{5/2}$, $^6H_{7/2}$ are magnetic dipole transitions, and $^4G_{5/2} \rightarrow {}^6H_{9/2}$ is an electric dipole transition [35]. Also, Sm^{3+} ion can be easily excited through the spin-allowed $^6H_{5/2} - {}^6P_{3/2}$

transition at about 407 nm. The $^4G_{5/2}/^4F_J$ low branching ratio exhibits the more intense visible transitions to the ^6H_J (J = 5/2, 7/2, and 11/2) [7], which shows that Sm^{3+} ion is an important activator for many different inorganic lattices to produce orange-red emission. Sm^{3+} ion can show both upconversion (UC) and DC mechanisms, in some host lattice [36–38].

2.6.3 Pr³⁺

Pr^{3+} ion shows the intra-4f narrow emission band due to 4f electrons, and it is well shielded by the outer 5s and 5p electrons from external forces, as discussed for Eu^{3+} ion [39]. The positions of the 4f energy levels are slightly dependent on the host matrix and are roughly the same as the free-ion levels. Usually, Pr^{3+} ions show a broadband excitation spectrum in the UV region covering 220–350 nm, and a series of sharp narrow excitation peaks appear in the visible region at 440–500 nm. Therefore, the narrow excitation band at 440–500 nm corresponds to the transition from ground state $^3H_4$ to excited state $^3P_2$, $^3P_1$, and $^3P_0$. Whereas the Pr^{3+} emission spectra consist of series of characteristic bands observed at 530–690 nm, which are attributed due to $^3P_1$–$^3H_5$, $^3P_0$–$^3H_5$, $^3P_0$–$^3H_6$, and $^3P_0$–$^3F_2$ transitions, these are assigned to f–f intra-configurational transitions of Pr^{3+} ion. Also, the Pr^{3+} ion excitation band is composed of multiplet ($^3H_6$ ^3Pn, $^1I_6$, with n = 0, 1, 2) in the visible region and the host charge-transfer (CT) transition in the UV region. The excitation edge is already weakly rising at about 360–370 nm and becomes very intense below 300 nm. It is well known that Pr^{3+} ions show photoluminescence under VUV excitation, and this has attracted much attention because of two possible emission processes: photon cascade emission (PCE) or f–d emission, which depends on the host lattice [40]. Therefore, Pr^{3+} ions have the unique feature that they emit light from the visible (VIS) to the infrared (IR) spectral regions, depending on the host matrix and the ion concentration.

2.6.4 Tb³⁺

Trivalent Tb^{3+} ion is also well characterized by the partially filled 4f shell and shielded by $5s^2$ and $5p^6$ orbitals. Photoluminesence (PL) emission bands of Tb^{3+} yield sharp lines in the optical spectra [41]. The application of RE ions depends on their "line-type" f–f transitions, which will be narrowed to the visible region, resulting in both a high efficiency and a high lumen equivalent. Therefore, there is an urgent need to find a stable, inorganic rare earth–based phosphor with high absorption in the near-UV region. Tb^{3+} ion with $4f^8$ configuration has complex energy levels and various possible transitions between 4f levels. Therefore, the transitions between these 4f levels are highly selective, and the lines are sharp in nature. The series of sharp excitation peaks between 250 and 500 nm corresponds to the Tb^{3+} intra-4f $4f^8$–$4f^8$ transitions. Also, the 4f–5d transition band of Tb^{3+} ion is in the UV region, and it can be detected in the excitation spectra. Thus, Tb-activated phosphors always show strong 4f–5d transition absorption bands around 200–300 nm. Usually, Tb^{3+} shows the characteristic blue and green emission peaks corresponding to Tb^{3+} intra-4f transitions from the excited levels to lower levels: $^5D_3 \rightarrow {}^7F_J$ (J = 2,3,4,5,6) and $^5D_4 \rightarrow {}^7F_J$ (J = 3,4,5,6) transitions, respectively. The PL emission spectra of Tb^{3+} ion show a completely different ratio between $^5D_3$ and $^5D_4$ emissions at lower and higher Tb^{3+} concentration. In this case, the excitation energy of an ion decaying from a highly excited state promotes a nearby ion from a lower state to the metastable state. Also, the energy gap between the $^5D_3$ and $^5D_4$ levels is close to that between the $^7F_6$ and $^7F_0$ levels. Thus, the characteristic emission of Tb^{3+} is observed in the visible and NIR regions in many host lattices.

2.6.5 Nd^{3+}

Nd^{3+} is a well-known efficient rare earth ion used in solid-state laser materials due to its intense emission at 1.06 mm [42–45]. From the emission properties, it is seen that Nd^{3+} could be a good candidate for UC of luminescent material in the visible light region [46, 47]. Therefore, the observed luminescence results have encouraged research into the photoluminescence properties of Nd^{3+} in the visible light region. However, there are very few reports available on the downconversion luminescence in Nd^{3+}-doped glasses or crystals in the visible light region. Another reason is that the luminescence efficiency of Nd^{3+} ion in many compounds is too low, and the absorption bands of Nd^{3+} correspond to transitions from the $^4I_{9/2}$ ground state to various excited levels. These transitions were assigned by comparing the band positions in the absorption spectra with a standard wavelength chart for the Nd^{3+} ion [48]. The various spectroscopic transitions observed are as follows: $^4F_{3/2}$ (877 nm), $^4F_{5/2} + {}^2H_{9/2}$ (803 nm), $^4S_{3/2} + {}^4F_{7/2}$ (745 nm), $^4F_{9/2}$ (684 nm), $^2H_{11/2}$ (630 nm) $^4G_{5/2} + {}^2G_{7/2}$ (583 nm), $^2K_{13/2} + {}^4G_{7/2}$ (526 nm), $^4G_{9/2}$ (519 nm)$\leftarrow {}^4I_{9/2}$, and moreover, the absorption bands corresponding to $^2D_{3/2}$, $^2G_{9/2}$, $^4G_{11/2}$, and $^2P_{1/2} \leftarrow {}^4I_{9/2}$ transitions.

2.6.6 Er^{3+}

Er^{3+} ion is one of the most important activators due to its UC luminescence process. Thus, Er^{3+} is also a more popular and efficient rare earth ion, employed due to its favorable energy level structure in the NIR spectral region corresponding to the two transitions $^4I_{15/2} \rightarrow {}^4I_{9/2}$ (800 nm) and $^4I_{15/2} \rightarrow {}^4I_{11/2}$ (980 nm), which can be efficiently excited, thereby producing a blue, green, and red UC emission [49, 50]. For the past few years, Er^{3+} ion has been widely considered due to its property of exhibiting three fluorescence transitions in blue, green, and red in the visible region [51–53]. Most of the research work has been carried out on the spectroscopic and laser characteristics of trivalent Er^{3+} ion in different host lattices. It is an excellent candidate for luminescence properties, because its metastable levels $^4I_{9/2}$ and $^4I_{11/2}$ can be conveniently populated by commercial low-cost high-power NIR laser diodes and optical devices [54, 55]. Also, the Er^{3+} emission bands observed at 524, 546, and 658 nm are assigned to the $^2H_{11/2} \rightarrow ^4I_{15/2}, ^4S_{3/2} \rightarrow ^4I_{15/2}$, and $^4F_{9/2} \rightarrow ^4I_{15/2}$ transitions, respectively. The multiphonon relaxation probability of the Er^{3+} ions depends on the energy gap between two successive levels and the phonon energy of the host lattice [56]. Thus, the lower the phonon energy of the host, the smaller the multiphonon relaxation probability. If we see the energy level diagram of Er^{3+} with possible UC mechanisms such as excited-state absorption (ESA) and energy transfer upconversion (ETU), at room temperature, no explicit dependence on the excitation wavelength can be observed for the green emission band in the NIR at 950–1000 nm for UC processes or at direct excitation in the visible spectral region ($^4I_{15/2} - ^4F_{7/2}$).

2.6.7 Yb^{3+}

Yb^{3+} is a well-known dopant due to its luminescence mechanism and the energy transfer process in the NIR region, and it has simple energy levels corresponding to $^2F_{5/2}$ and $2F_{7/2}$ levels, respectively, separated by a difference of about 10,150 cm^{-1}. Therefore, this impurity ion is very suitable for emitting light in the NIR region at around 985 nm. It consists of only two energy levels, it has no branching, and the emission originates from $^2F_{5/2} \rightarrow ^2F_{7/2}$ transition. Thus, it is helpful to convert NIR light into visible light by a UC process when doped into a host lattice [57–61]. Two Yb^{3+} ions in the excited state can transfer their energy to an activator, changing it to the excited state, and it then gives an emission in the visible region. Yb^{3+} has a long lifetime in the excited state, and thus, the possibilities of the cooperative processes are significant. In the DC or quantum cutting process, the energy of the ions is transferred simultaneously from a sensitizer ion in the excited state to two Yb^{3+} ions, which is feasible for the same reasons [62]. Yb^{3+} can be introduced into a host crystal at high concentration without noticeable concentration quenching [63], which is very important for efficient excited state energy transfer. The NIR emission in Yb^{3+} ion is also quite important for the energy conversion process in a solar cell [64]. The forbidden transition assigned to $^2F_{7/2} \rightarrow ^2F_{5/2}$ of Yb^{3+} is very weak. Yb^{3+} host lattices have potential technological applications in the field of display and related technology. Also, they can generate tunable laser light in the infrared region from 920 to 1060 nm and visible blue-green emission at about 500 nm by means of cooperative effects [65].

2.6.8 Ho^{3+}

Ho^{3+} ion shows emission bands from 0.55 to 4.9 μm. Laser emissions observed at 1.2 μm ($^5I_6 \rightarrow ^5I_8$) and 2.0 μm ($^5I_7 \rightarrow ^5I_8$) have been outputted successfully from activator Ho^{3+} when doped into suitable host materials [66, 67]. The mid-IR wavelength emission is especially useful for laser radar imaging because of high atmospheric transmittance and low background noise [68]. Ho^{3+} ion shows spin-forbidden transitions from $4f^9$ 5d excited state to the ground state. The emission bands are assigned to spin-allowed and spin-forbidden transitions between the $4f^9$ 5d high-spin (HS), $4f^9$ 5d low-spin (LS), and $4f^{10}$ electronic configurations of the Ho^{3+} ion. Among the different lanthanide ions, when Ho^{3+} is doped into glass matrices, it exhibits intense emission corresponding to the UV, visible, and NIR regions. Thus, it is one of the most important activator ions applicable as a chromophore for IR lasers due to its favorable energy level structure. Trivalent holmium ion forms several metastable states, which offer multiple possibilities of laser transitions in the longer wavelength region range, from the UV region up to IR. Also when it is doped with inorganic host glasses then it shows the laser emission with new multi-energy level schemes [69–70]. Thus, it is a promising ion for the UC mechanism due to its specific energy level arrangements, which allow a potential UC process when activated by host glasses with low-energy photons. Another advantage of this ion is that it shows several energy levels that are not substantially disturbed by multiphonon decay. Hence, it is used for UC and ESA processes. Therefore, this transition consequently populates the higher energy levels [71] and originates the broad fluorescent emission band. Due to the presence of UC capability from IR to visible light, Ho^{3+}-doped glasses are being extensively used in many applications, such as visible lasers, optical data storage systems, sensors, medical diagnostics, solar cells and so on [72]. Also, it is applicable to three-dimensional (3D) displays, TV, and 3D cameras. With increasing demand for digital optical communication, it is a challenging task for researchers to develop amplifiers in the 800 nm range due to its advantages in the optical communication window. Ho^{3+}-doped host matrixes seem to be better candidates for this purpose due to their multiple emission characteristics, such as 2 µm emission, visible wavelength UC, and so on.

2.6.9 Tm^{3+}

Tm^{3+} ion with metastable levels is suitable to serve as the active center for a tunable solid-state laser with UC pumping [73–76]. Among the different oxide-based host materials, aluminates doped with Tm^{3+} ion are excellent laser hosts due to their good mechanical strength and thermal stability [77–79]. Also, Tm^{3+} ions show highly efficient UV and blue emission. The emission band of Tm^{3+} corresponds to the well-known $^3F_4$–$^3H_6$ electrical transition, and it is used for fabricating a 2 µm light source. The broad emission band at 2 µm is suitable for a tunable laser source. Usually, a Tm^{3+}-doped inorganic host lattice shows a blue-emitting performance due to the blue transition of $^1D_2 \rightarrow {}^3F_4$ under near-UV light excitation [80–82]. The multiphonon nonradiative transitions in Tm^{3+}-doped oxide hosts are weak, and most of the transitions have high quantum efficiency, so that many laser actions in crystal doped with Tm^{3+} have been reported in the middle IR, NIR, visible-, and near-UV range. Such long wavelength emission, covering the UV to NIR range, is important to investigate novel Tm^{3+}-doped phosphors for possible applications in industry, especially in the field of solid-state lighting as well as in solar cell applications.

2.7 Energy Transfer in Rare Earth Ions

In many luminescent materials, the excitation energy is not absorbed by the activator but goes elsewhere; that is, another impurity ion, which absorbs the exciting radiation (called a *sensitizer*), is added to the host lattice, and afterward transfers the energy to the activator, as shown in Figure 2.6. In Figure 2.7, the transition from S→S* shows excitation, while transitioning from A_2^*→A shows emission. Level A_2^* is populated by energy transfer. Hence, it decays nonradiatively to the lower level A_1^*. This can prevent the possibility of reverse energy transfer. Therefore, energy transfer is the process by which the excited state energy is transferred from the donor to an activator ion.

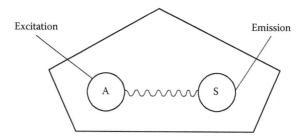

FIGURE 2.6
Energy transfer from sensitizer S to an activator A ion.

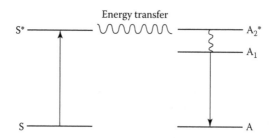

FIGURE 2.7
Energy transfer from S to A.

This energy transfer occurs without the emergence of a photon as a result of the multiple interactions between donor and activator impurity ions. When the energy difference between the lower and higher states of a sensitizer and an activator are equal, or if a suitable electrostatic or magnetic interaction takes place between them, the process of energy transfer takes place. Also, if spectral overlap exists between the photoluminescence emission spectrum of the sensitizer ion and the absorption spectrum of an activator ion, this gives strong evidence of the possibility of energy transfer. Energy transfer was studied by Foster and Dexter. They predicted that the rate of energy transfer will be proportional to the overlap of the donor emission and the acceptor absorption spectra and to R^{-6}, where R is the donor-to-acceptor distance. Generally, an energy transfer process can be represented by

$$S^* + A \rightarrow S + A^*, \text{ where the asterisk indicates the excited state.}$$

Energy transfer between ions has important applications in sensitizing solid-state lasers, infrared quantum counters, and infrared to visible converters, and is also useful in the fabrication of tunable LEDs [83, 84].

2.7.1 Theory of Energy Transfer

The process by which the excitation of a certain ion migrates to another ion is called *energy transfer*. It plays an important role in the development of high-energy-efficient luminescent materials to develop new lighting and display devices. The basic mechanisms involved in energy transfer processes between ions can be stated as follows [55]:

- Resonant radiative energy transfer through emission of sensitizer and reabsorption by the activator.
- The nonradiative transfer associated with resonance between sensitizer and activator ions.
- CR occurs between two identical ions.
- Multiphonon-assisted energy transfer.

The energy transfer mechanism between two identical ions is shown in Figure 2.8 [85]. From Figure 2.8a, it is seen that the effective radiative energy transfer depends on how well the activator ion's luminescence is excited by the sensitizer emission. Thus, the significant condition of spectral overlap between the emission spectrum of the sensitizer and the absorption spectrum of the activator is required. If primarily radiative energy transfer takes place, the decay time of sensitizer luminescence is not affected by an increase in the concentration of the activator ions. The nonradioactive energy transfer is shown in Figure 2.8b, in which there exists a considerable decrease

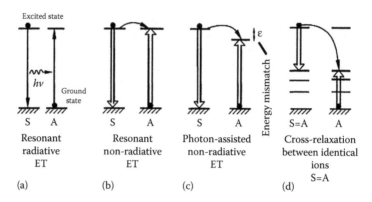

FIGURE 2.8
Energy transfer mechanism between two ions. (a) Resonant Radiative Energy Transfer (b) Resonant Nonradiative Energy Transfer (c) Photon-Assisted Nonradiative Energy Transfer (d) Cross-Relaxation between identical ions S=A (From Nalwa, H.S. and Rohwer, L.S., *Handbook of Luminescence, Display Materials, and Devices*, American Scientific Publishers, California, 2003.)

in decay time of the sensitizer luminescence with the increase in activator concentration. If the energy difference between the ground and excited states of the sensitizer is equal to that of the activator, there exists a suitable interaction between systems, and hence, the process of energy transfer takes place [86]. There are two types of interaction for resonant energy transfer, as illustrated in Figure 2.9a. For the electrostatic interaction between donor (D) and acceptor (A), a multipole interaction takes place, which can be classified as dipole–dipole (d–d), dipole–quadrupole (d–q), quadrupole–quadrupole (q–q) interaction. The transfer probability, PDA, between D and A can be represented by Equation 2.5 [87]:

$$PDA = \frac{2\pi}{\hbar} \left| < D^*, A \left| HDA \right| D, A^* > \right|^2 \int gD(E) \cdot gA(E) \, dE \qquad [2.5]$$

where:

$<D^*, A\|$ and $\|D, A^*>$	represent initial state and final state, respectively
HDA	is the Hamiltonian of interaction
$\hbar$	is the Planck constant
$\int gD(E) \cdot gA(E) \, dE$	denotes the integral of spectral overlap between D and A
For d–d, d–q and q–q	interactions, transfer probability is inversely proportional to R^6, R^8, and R^{10}, respectively.

Figure 2.9b indicates the quantum mechanical exchange interaction, in which D and A are closer to each other, and their wave functions are overlapped; therefore, the energy transfer mechanism from D to A can be

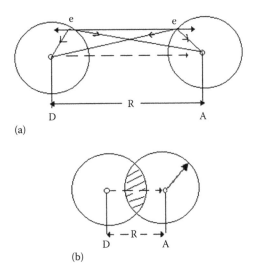

FIGURE 2.9
(a) Electrostatic interaction; (b) quantum mechanical exchange interaction. (From Li, Y.Q., et al., *Chem. Mater.*, 17, 3242, 2005.) [87]

represented by exchange interaction. The transfer probability P_{ex} (R) can be expressed by Equation 2.6 [87]:

$$Pex(R) = \frac{2\pi}{\hbar} K^2 \exp\frac{-2R}{L} \int FD(E) \cdot FA(E) \, dE \qquad [2.6]$$

where:
   K^2 is a constant
   R is the distance between D and A
   L denotes the average radius of the excited state (D) and ground state (A), and the integrated item represents spectral overlap

The energy transfer mechanisms in phosphors are the outcome of many competitive interactions, which prominently depend on the characteristic transition of D and A, and the corresponding distance between them. If two ions have different excited states, the probability of energy transfer should drop to zero, and the overlap integral gD(E)gA(E) dE also becomes zero [88]. On the other hand, it is found that the energy transfer can take place without phonon-broadened electronic overlap, which suggests that overall energy conservation is maintained. The CR process between ions refers to all types of DC ET between identical rare earth ions. In such cases, there exists the same kind of ion, which works as both sensitizer and activator. CR may give rise to a diffusion process between sensitizers when the levels involved are identical or to self-quenching when levels are different. In the first case, there is no loss of energy, whereas in the second case, the loss of energy takes place.

2.7.2 Excitation by Energy Transfer

Strong excitation can frequently be achieved by energy transfer. Therefore, energy transfer results either in the enhancement or in the quenching of emission. The energy transfer from a host crystal to activators leads to host excited luminescence. The radiation emitted by a phosphor originates from a luminescence center (activator) which is doped into the host lattice. The activator can be excited directly, in which case the activator itself may absorb the excitation energy. However, in some cases, the excitation energy is not absorbed by the activator itself but in the other center; it then transfers the energy to the activator. The center that absorbs the radiation is called the *sensitizer* (S), and the center to which the energy is transferred is called an *activator* (A).

There is no fundamental difference observed between A and S. The luminescent center can play the role of S or A depending on the nature of the host lattice. If a center S absorbs a quantum of radiation, four things can happen. The excitation radiation (exc_s) is absorbed by a center S. This process is followed by one or more processes: (a) emission from S (emn_s) (probability P_s^r); (b) nonradiative loss from S by heat dissipation (probability P_s^{nr}); (c) transfer of energy to another center of type S (P_{SS}); (d) transfer of energy to another center of type A (P_{SA}). These processes are depicted in Figure 2.10.

1. S shows luminescence itself.
2. S returns nonradiatively to the ground state.
3. S transfers its excitation energy to A.
4. S transfers its energy to another center S.

The energy transfer in a phosphor can be determined by measuring the absorption and emission spectra of S and A and the decay time of the luminescence from S as a function of the concentration of A.

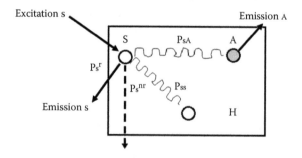

FIGURE 2.10
Energy transfer from sensitizer S to the activator A.

2.8 Photoluminescence

Photoluminescence is a type of luminescence process in which a photon acts as the excitation source. The electronic states of solids are excited by some incident light energy, which always occurs in the form of short wavelength, usually in the near-UV range. The mechanism of absorption and emission of light is shown in Figure 2.11; if a light particle (photon) has energy greater than the band gap energy, it can be absorbed, and by this means, it raises an electron from the valence band up to the conduction band, crossing the forbidden energy gap. In this photoexcitation process, the electron generally has excess energy, which is lost before it arrives at the lower energy level, that is, the conduction band. As it falls down, the energy it loses is converted back into a luminescent photon, which is emitted from the material. This emission occurs by photon excitation. Hence, in the photoluminescence process, the light is absorbed by materials for a significant time, and later, they produce light of a frequency that is different from the frequency of the absorbed light [6].

2.8.1 Rare Earth–Activated Phosphor

At present, many rare earth–activated inorganic host materials have been reported by researchers for possible application in SSL technology. But only a few phosphor hosts are commercially used in phosphor industries. Among the other polymer composite and organic hybrid materials, rare earth–activated inorganic phosphors are more attractive due to their long lifetime, thermal stability, and environmental and heat resistance capabilities. Table 2.3 shows the different families of the inorganic materials, such as borates, fluorides, silicates, sulfates, sulfides, titanates, titanates, vanadates, phosphates, and tungstates, out of which borates, vanadates, phosphates, and tungstates are considered to be excellent hosts to study upconversion/downconversion (UP/DC) mechanisms when doped with Er^{3+}, Yb^{3+}, Eu^{3+}, Nd^{3+}, and so on.

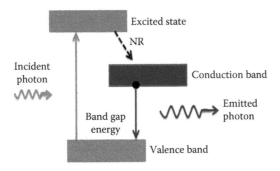

FIGURE 2.11
Photoluminescence excitation and emission process.

TABLE 2.3

Types of Phosphor Host Used in Display Devices

Phosphor Host	Different Types of Phosphor Host Used in Display Devices
Borates	$Li_2B_4O_7$:Cu^{2+}, MgB_4O_7:Dy^{3+}, YBO_3:Eu^{2+}, SrB_4O_7:Sm^{2+}, LaB_3O_6:Bi, Gd, $Sr_2B_5O_9Br$:Ce, $InBO_3$:Tb^{3+}, (La, Ce, Gd) MgB_5O_{10}:Tb^{3+}
Fluorides	LiF:Mg,Ti, LiF:Mg,Cu,P, CaF_2:Dy, $LiSrAlF_6$:Cr, $LiBaF_3$, $KMgF_3$, $LiCaAlF_6$:Ce^{3+}
Silicates	$BaSi_2O_5$:Pb, Zn_2SiO_4:Mn, Y_2SiO_5:Ce, Mg_2SiO_4:Tb
Sulfates	$CaSO_4$:Dy, $CaSO_4$:Tm, $CaSO_4$:Eu^{2+}, $CaSO_4$:Eu^{3+}, $BaSO_4$:Eu^{2+}
Sulfides	ZnS:Ag, ZnS:Ag,Cl, $SrGa_2S_4$:Eu^{2+}, CaS:Eu,Sm, Gd_2O_2S:Tb^{3+}, Y_2O_2S:Eu^{3+}, Zn-CdS:Cu
Titanates	$SrTiO_3$:Al,Pr^{3+}, $CaTiO_3$:Al,Pr^{3+}
Vanadates	YVO_4:Eu^{3+}, Y(V, P)O_4:Dy^{3+}, Y(V, P)O_4:Eu^{3+}
Phosphates	$Sr_5(PO_4)_3F$:Mn, $Sr_5(PO_4)_3$(F, Cl):Sb,Mn, $Sr_5(PO_4)_3Cl$:Eu^{2+}
Tungstates	Cd(WO4):Yb^{3+}, Sr_2MgWO_6:Nb,Ta^{3+}

2.9 Classification of Materials

A material can be defined as a chemical substance or mixture of substances
that is obtained in solid form or in the condensed phase, and which is
intended to be used for many engineering applications. Materials are classified on the basis of their properties, such as physical, chemical, geological, and in some cases, biological [89]. Materials science is an important and
interdisciplinary area of research in the branch of pure and applied sciences,
such as physics, chemistry, electronics, environmental engineering, and
nanotechnology, because of its growing importance in global development.
In general, a solid can be defined as a regular and periodic arrangement
of points in three-dimensional space. The applications of the materials are
also based on their stability, structure, phase, density, melting point, and so
on. They are divided into two classes: crystalline and non-crystalline. Some
examples of traditional materials are metals, semiconductors, ceramics, and
polymers [89]. Recently used novel and advanced materials include nanomaterials, biomaterials, and so on [89]. A short description of the types of
materials and their properties, such as physical, chemical, mechanical, and
optical, is given in the following section.

2.9.1 Metals

These consist of metallic elements as given in the periodic table. They have
a specific atomic number including a stable oxidation state, having a large
number of nonlocalized electrons. The properties of the metal are directly
attributable to the unbonded electron. Usually, metals are discussed based on

their metallic properties, such as luster, opacity, malleability, ductility, melting point, electrical and thermal conductivity, and so on. An alloy is a mixture of two or more elements in which the host component is a metal. Most pure metals are either too soft, brittle, or chemically reactive. Thus, combining different ratios of metals as alloys can be helpful to modify the properties of pure metals and to obtain desirable characteristics. Some well-known examples of metallic materials are aluminum (Al^{3+}), copper (Cu^+), zinc (Zn^{2+}), and so on, and their alloys. Generally, they are available in bulk or powder form. We know that metals have higher densities than most non-metals [90]. Groups IA and IIA correspond to alkali and alkaline earth metals. They are referred as the *light metals* due to their low density, low hardness, low melting points, and so on [90]. The high density of most metals corresponds to the tightly packed crystal lattice of the metallic structure.

2.9.2 Ceramics

Ceramics are intermediate compounds between metallic and non-metallic elements. Materials such as oxides, nitrides, and carbides are types of ceramics. Recently, graphene has been most widely used in the synthesis of many host materials to improve the existing properties of materials, and it falls into the category of ceramics. Today, ceramic materials are widely used in many applications. They fall within the given classification, and they are composed of clay minerals, cement, and glass. Here, glass also falls into the class of ceramics, which is amorphous in nature, so that the atoms are randomly oriented in all directions. They are characterized by standard laboratory tests such as hardness, brittleness, resistance, conductivity, and so on. The physical properties of any ceramic substance are associated with its crystalline nature and chemical composition. The solid-state and structural chemistry of ceramics defines the fundamental relation between microstructure and properties, including localized density, grain size distribution, type of porosity and second-phase content, mechanical strength, hardness, toughness, dielectric constant, and optical properties exhibited by transparent ceramic materials. Some semiconductor materials come under the class of ceramics. Ceramics have been especially applicable in many electronic devices and electrical components for the past few decades.

2.9.3 Polymers

A polymer is composed of a large molecule or macromolecule formed by many repeated subunits. Therefore, the characterization of polymers is one of the most important aspects of polymer science and technology, which need to discuss their properties and applications. It is difficult to characterize polymers as easily as other categories of materials, due to their long-chain, coiled, and entangled structure. The presence of structural defects, such as the formation of short or long branches or different chain lengths, also sometimes causes complexity in structure, and hence, it becomes difficult to know the exact structure of the polymer. Polymers are characterized by conventional techniques such as Fourier transform infrared spectroscopy (FTIR), UV-visible spectroscopy, nuclear magnetic resonance (NMR), gas chromatography–mass spectrometry (GC–MS), elemental analysis, and so on. Polymers include familiar synthetic plastics such as polystyrene. Polymers such as poly(methyl methacrylate) (PMMA) and hydroxyethyl methacrylate-methyl methacrylate (HEMA:MMA) are used in solid-state dye lasers due to their high surface quality and transparency. Recently organic polymer-based materials have become very much useful in *organic light-emitting diodes* (OLEDs) and solar cells due to their extensive color quality and light conversion efficiency. Lasers made up of organic materials show a very narrow linewidth, which is useful for spectroscopy and analytical applications [91, 92]. Both synthetic and natural polymers are prepared via the process of polymerization, which is also a combination of many small molecules, called *monomers*. Their large molecular mass relative to small-molecule compounds produces unique physical properties, such as toughness and viscoelastic properties, and they always tend to form glasses and semi-crystalline structures rather than crystals.

2.9.4 Composites

A composite is formed by the combination of two or more constituent materials differing in their physical and chemical properties. When the final product is produced, it is different from the individual components. Recently, many researchers have also begun to actively include sensing, actuation, computation, and communication into composites [93], which are also called *robotic materials*. Some typical examples of composite materials are wood, clad metals, fiberglass, reinforced plastics, cemented carbides, and so on. Among these, fiberglass, in which glass fibers are embedded within a polymeric material, is mostly used in many applications. In this composite, the fiberglass acquires strength and flexibility from the glass and from the polymer. Many efforts have been made by researchers to develop new composite materials with desired properties. Thus, recent developments in the material field have involved composite materials. Also, nanocomposites are a rapidly expanding field for the development of science and advanced technology due to the availability of new advanced materials with unique electrical and optical properties. The class of nanocomposites includes organic or inorganic new hybrid materials, which is a fast-growing field of research. Continuous efforts have been made to improve the properties of nanocomposite and to control their morphology and interfacial characteristics. Polymer-based nanocomposites have many important applications in electronics as charge storage capacitors, integrated circuits (ICs), and electrolytes for solid-state batteries.

2.9.5 Twenty-First-Century Materials

2.9.5.1 Smart Materials

Today, smart materials are receiving great interest in all the fields of science and technology because of their numerous applications in many areas, such as research, engineering, medical, nanotechnology, and so on. Following recent developments in the field of materials science and technology, many new advanced, high-quality, cost-effective materials have been coming into use in various fields of engineering and technology. In the past few decades some new materials were introduced which exhibit multifunctional properties, therefore for detail analysis they required some different characterization techniques to know their behavior at micro and nanoscale. With continuous evolution in the research field, in the last few years, the concept has moved toward composite materials, and currently, the next evolutionary step deals with the concept of smart materials and their potential applications in low-cost and environment-friendly technology. This is an era of new-generation materials, including conventional structural and advanced functional materials. In 1988, Rogers et al. explained that smart materials are materials that have the ability to change their physical properties in a specific manner with respect to specific stimulus inputs such as pressure, temperature, electric and magnetic fields, chemicals, hydrostatic pressure, nuclear radiation, and so on. The variable physical properties are shape, stiffness, viscosity, and damping. Therefore, the use of smart materials in the twenty-first century has many advantages for environmental sustainability; for example, it may respond to environmental changes with optimum conditions and functions according to the environment. The smartness of the materials indicates the self-adaptability, self-sensing, memory, and multiple functionalities of the materials or structures. Thus, there are many applications of smart materials in the fields of science and engineering due to their varied response (e.g., aerospace, civil engineering applications, energy materials, etc.). Therefore, today, there is an urgent demand for smart materials that have the potential to solve engineering problems and provide an opportunity for the manufacture of new products that will generate revenue for sustainable development.

Some examples of smart materials are

- Piezoelectric materials
- Shape-memory alloys and shape-memory polymers
- Magnetostrictive materials
- Magnetic shape memory
- Smart inorganic polymers
- pH-sensitive polymers
- Temperature-responsive polymers

- Chromogenic systems
- Photomechanical materials
- Polycaprolactone
- Self-healing materials
- Dielectric elastomers
- Magnetocaloric materials
- Thermoelectric materials

2.9.5.2 Nanomaterials

Nanomaterials are defined as solids that have a crystallite size in the range from 1 to 1000 nm (10^{-9} m) but usually 1–100 nm [94]. Nanomaterials research takes a materials science–based approach to understanding physical, chemical, and size-dependent properties. There are well-known techniques to synthesize nanoparticles of different sizes and shapes, such as sol-gel and combustion techniques. Materials with nanoscale structure often have unique optical, electronic, or mechanical properties [95]. Nowadays, rare earth–doped nanomaterials have received a lot of interest due to their high surface to volume ratio and quantum confinement effect. The optical properties of nanomaterials in the form of nanoparticles, nanorods, nanowires, nanotubes, and colloidal or bulk nanocrystals are of great interest not only for basic research but also for interesting applications in engineering [96–99]. Nanomaterials have different applications from those of the bulk materials. More effort has been devoted to the development of nanomaterials, which have been applied in many areas, such as hybrid solar cells [100], transparent conductive films [101], gas sensors [102], and nanophosphors for solid-state lighting, especially used in LEDs [103]. Among these, quantum dots, nanotubes, nanowires, and nanorods have received much attention due to their interesting optical and electrical properties, applicable in optoelectronic devices [104]. These properties are dependent on their size and shape, which can be changed by synthetic techniques. Organic-based nanomaterials may also be applicable in optoelectronic devices such as organic solar cells, OLEDs, and so on. These devices work on the principle governed by photoinduced processes, such as electron transfer and the energy transfer process. The working performance of the devices is based on the efficiency of the photoinduced process. Therefore, it is important to understand photoinduced processes in organic/inorganic nanomaterial composites to use them in organic optoelectronic devices. Also, nanoparticles made of metals, semiconductors, or oxide-based materials are of great interest due to their extensive mechanical, electrical, magnetic, optical, and chemical properties [105, 106]. These nanoparticles have tremendous applications in dye degradation from wastewater (e.g., TiO_2 nanoparticles). Hence, nanoparticles are a bridge between bulk materials and atomic or molecular structures. Rare earth–doped nanocrystalline materials have received considerable attention due to their various applications in flat panel displays, lasers, solar cells, optics, three-dimensional display technologies, optical telecommunication, infrared quantum counters, sensors, and bio-imaging [107–112].

2.10 Applications of Phosphors

For the last few decades, rare earth–doped phosphors have been used in many display and lighting devices. They play an important role in converting high-energy incident photons into low-energy photons; that is, the conversion of photons takes place from the short wavelength region to the long wavelength region. Therefore, phosphors have different color emission characteristics, and these can be varied by doping a suitable rare earth ion in the phosphor host. The host materials support the emission centers. In general illumination and fluorescent lighting devices, phosphor materials convert incident UV radiation into visible light. Traditional lamps made from tungsten filament are not very energy efficient; they consume more power and generate heat in the atmosphere, while in discharge lamps such as Hg-vapor lamps, around 60% of the energy is converted to light [113]. Therefore, considering this drawback of traditional lighting, today, there is more demand for environment-friendly and energy-efficient phosphors to develop low-cost solid-state lighting devices [114]. Hence, there is an urgent need to develop novel, highly efficient phosphor materials to meet the demand from the lamp industry. In the following section, we will discuss in detail the major advantages of phosphors in CFLs, LEDs, and display devices.

2.10.1 Fluorescent Lamps

A fluorescent lamp is a very efficient generator of ultraviolet energy at a wavelength of mostly 254 nm and sometimes 185 nm. The fluorescent lamp phosphors absorb the ultraviolet radiation and convert it into visible light. In the lamp industry, the term *light* refers to light in the ultraviolet to near infrared regions. In a fluorescent lamp, a very fine powder form of the phosphor is applied as a uniform coating on the inner surface of the glass tube, as shown in Figure 2.12. The phosphor used for coating must be in a powder form. The phosphor powder is transformed into a paint, which is then applied to the upper interior of the vertical lamp tube and drained in such a manner as to form a very uniform coating. The phosphor coating is then dried and heated near to the melting point of the glass to burn off the organic components of the paint. To improve the adhesive properties of the phosphors, usually, borates or very fine aluminum oxide may be added as binders to the paint.

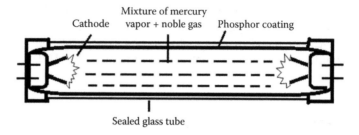

FIGURE 2.12
Internal surface of the mercury lamp.

This also prevents chipping of the coating. Table 2.4 shows some of the phosphors used in fluorescent lamps. These available phosphors can be blended to produce white light. Calcium halophosphate is the dominant phosphor used in the lamp industry [115, 116], because fluorescent lamps fabricated using this phosphor have better outputs and color temperatures compared with lamps fabricated using zinc beryllium silicate and magnesium tungstate materials.

2.10.2 Light-Emitting Diodes (LEDs)

During the 1970s, Nixie tubes were replaced by LEDs, and afterward, LEDs found applications in several display devices. The emission from LEDs is observed in the visible region. The ideal phosphor for use in LEDs should have a band gap always greater than 1.8 eV, and it should be compatible for doping with n- or p-type materials for the formation of a competent homojunction. LEDs show good luminescence efficiency, color quality, and cold light as compared to other lighting sources. An In_xAl_{1-x}-based semiconductor system had been studied previously for use in optical devices [117, 118]. The innovation of GaN-based blue LEDs [119] and the metal-organic chemical vapor deposition (MOCVD) technique used for LED chip production have modernized the field of solid-state lighting.

TABLE 2.4

Some of the Phosphors Used in Fluorescent Lamps

Name	Formula	Activator	Emission Color
Zinc silicate	Zn_2SiO_4	Mn	Green
Zinc beryllium silicate	$(Zn, Be)_2SiO_4$	Mn	Orange
Calcium silicate	$CaO{:}SiO_2$	Pb, Mn	Salmon
Cadmium borate	$CdO{:}B_2O_3$	Mn	Pink
Calcium tungstate	$CaWO_4$	Pb	Blue
Magnesium tungstate	$MgWO_4$	Self	Blue
Calcium halophosphate	$Ca_{10}(PO_4)_6(F, Cl)_2$	Sb, Mn	White

The emission spectrum of GaN LEDs has effectively been extended into the green spectral region with still higher efficiencies by varying the content of the InGaN active layer. Therefore, it is possible to produce white light by direct conversion from an electrical current in LEDs with a better efficiency compared with halogen lamps. LEDs emitting at 370 nm, coupled with RGB phosphors (having efficient absorption at 370 nm), can provide white light with good CRI. The search for stable inorganic rare earth–activated phosphors with high absorption in the UV/blue spectral region has, therefore, became an interesting field of research. White LEDs are usually composed of blue InGaN LEDs coupled with a coating of a suitable phosphor material such as cerium (III)-doped YAG (YAG:Ce^{3+} or $Y_3Al_5O_{12}$:Ce^{3+}), that is, yellow phosphor. In this case, the phosphor materials absorb the light from the blue LED chip and show the white light emission; the white band includes all colors of wavelength from violet to red. Some rare earth–doped SiAlONs are also used in display devices. One such material is Eu^{2+}-activated β-SiAlON, which absorbs ultraviolet light and converts it into broadband visible emission. But the quality of light does not change significantly with increase in temperature; hence, it is a more thermally stable material for use in LEDs. Also, it is used as a potential green-emitting DC phosphor candidate for solar cells and also in white LEDs. Generally, white LEDs are fabricated by the combination of a blue LED coupled with a yellow phosphor or with the combination of green and yellow SiAlON phosphor and a red-emitting $CaAlSiN_3$-based (CASN) phosphor [120–122]. White LEDs are fabricated by coating a mixture of RGB phosphor on a near ultraviolet-emitting LED chip. The quality of light depends on the combination of phosphors (such as RGB or BY) coupled with the semiconductor chip.

2.10.3 Flat Panel Displays (FPDs)

Flat panel displays, as shown in Figure 2.13, include a growing number of technologies, such as video displays that are much thinner and lighter than traditional television and video displays that use cathode ray tubes. They are usually less than 100 mm thick. FPDs require a small amount of power to accelerate the electrons from the cathode to the anode. They are defined as ideal displays because they are thin, have an even surface and low volume, have high resolution and contrast, and are lightweight. They are used in many modern portable devices, such as laptops, cellular phones, and digital cameras [123].

Some examples of FPDs are

- Plasma display panels (PDPs)
- Liquid crystal displays (LCDs)
- Organic light-emitting diode displays (OLEDs)
- Electroluminescent displays (ELDs)
- Surface-conduction electron-emitter displays (SEDs)
- Field emission displays (FEDs)

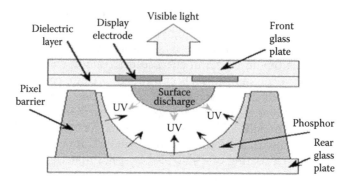

FIGURE 2.13
Plasma display panel.

2.10.4 Cathode Ray Tubes (CRTs)

A CRT consists of a vacuum tube containing an electron gun (cathode) and a phosphor-coated screen, as shown in Figure 2.14. The cathode is used as a filament, which on heating produces the electron beam. Electrons pass in a straight line through the anode and focusing coils, which are positioned on the upper and lower sides of the tube. The electrons are then focused and accelerated through the tube. Finally, they are incident on the fluorescent screen, showing a bright spot on the screen surface. Copper windings are wrapped around the tube and act as steering coils. These coils create magnetic fields inside the tube, and these magnetic fields steer the beam toward the screen. By varying the voltages in the coils, the electron beam can be positioned at any point on the screen. Color CRTs have three electron guns, one for each primary color. CRTs are used in oscilloscopes, television and

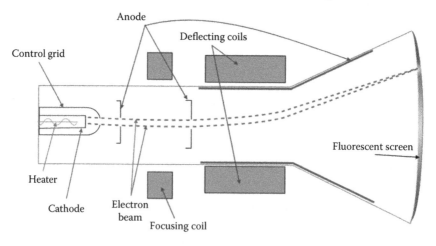

FIGURE 2.14
Cathode ray tube (commons.wikimedia.org).

computer monitors, and radar targets. Typical values of cathode to anode distance range are observed between 25 and 100 cm. CRTs are very bulky, and when bigger screens are required, the length of the tube must increase [124, 125]. For this application, the phosphor material should have specific characteristics [126, 127]. The application of phosphor powders for the purpose of screen coating is dependent on various interrelated parameters such as screening method, particle size distribution, and particle shape [128]. Some phosphors used in CRTs are as follows.

- Red phosphor: Yttrium oxide is used as the red phosphor in color CRTs.
- White phosphor: The mix of $ZnS:Ag^+$ and $CdS:Ag$ is the white P_4 phosphor used in black and white television CRTs, and YVO_4, Eu^{3+} is a primary red color used in television (1964) [129, 130].
- Green phosphor: This is the combination of zinc sulfide with copper, which gives green light peaking at 531 nm with long afterglow emission.
- Blue: $ZnS:Ag$ is also one of the most efficient blue phosphors used in color CRTs.

2.10.5 PV Technology

PV solar cells are appealing and promising renewable power sources due to some specific advantages; for example, during operation they do not generate pollution and greenhouse gas emissions. But the current PV technology suffers from some major disadvantages, such as low conversion efficiency, because the power output is dependent on incident sunlight, and they do not capture all wavelengths of incident light. Enhancing the efficiency of a PV solar cell beyond the Shockley–Queisser limit is only possible with the help of a DC or UC phosphor, which helps to fill the spectral mismatch in energy conversion. The working mechanism of such an energy conversion layer is illustrated in Figure 2.15. Only the incoming solar spectrum is modified to generate more power output from the PV cell. Therefore, adding such a layer to the device structure is feasible for many existing solar technologies. Hence, considering such advances in technology, we can reduce the cost and enhance the efficiency of PV devices by developing some new techniques.

2.10.6 Upconversion Phosphor

The solar spectrum can be modified by employing a UC mechanism. Normally, in a single–band gap solar cell, photons with energy lower than the band gap of the solar cell are lost. For example, a solar cell with a band gap of 2.0 eV loses approximately 50% of the total energy from the solar spectrum. Therefore, in a UC process, two or more low-energy photons

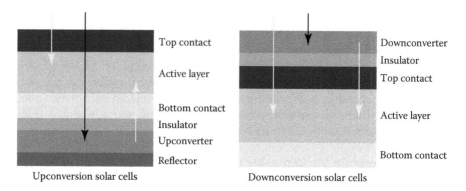

FIGURE 2.15
Upconversion and downconversion process in solar cell.

(sub–band gap photons) can be transferred into one high-energy photon (i.e., with energy exceeding the band gap). By incorporating the layer of UC materials on the front surface of the solar cell, more low-energy photons can be harvested, which results in increased current, and the efficiency of the solar cell increases. The final result of UC in a solar cell is similar to that of the intermediate band solar cell. In both types of cell, a parallel combination of a one-step absorption/charge generation process and a two-step absorption/charge generation process is in operation. The difference between them is that in the intermediate band solar cell, all processes are taking place inside the photovoltaic active layer, making the realization of an efficient device a formidable challenge, while in the UC solar cell, the upconverter layer can be electronically isolated from the active cell and located behind it. Therefore, in UC solar cells, the active cell and the UC layer can be optimized separately to gain optimum efficiency of the solar cell.

2.10.7 Downconversion Phosphor

Downconversion is also known as the *quantum cutting process* [131]. In this process, one incident high-energy photon is split by a material into two or more lower-energy photons. When the energy of these secondary photons is better matched with the band gap of the active material, it is absorbed by the active material and increases the current generated from the solar cell, while the voltage remains the same. Hence, the efficiency of the solar cell is increased. Therefore, the use of such a DC phosphor layer on the front of the solar cell is an ideal approach to cover the spectral mismatch in the solar cell. Trupke et al. calculated the optimum efficiency of a DC solar cell as a function of the band gap of the active material layer [132]. According to their theoretical calculation, a maximum conversion efficiency of up to 39.6% can be reached under a 6000 K blackbody spectrum and when the band gap of the active semiconductor layer is 1.05 eV. This optimum band gap is very close to the band gap of crystalline silicon (1.12 eV), which indicates that DC is a promising way to increase the efficiency of existing (multi)crystalline silicon solar cells. $LiGdF_4:Eu^{3+}$ and $GdF_3:Eu^{3+}$, BaF_2: Gd, Eu phosphors have been reported in the literature as active DC phosphors for solar cell applications.

2.11 Types of Host Material

2.11.1 Tungstates

A tungstate is a compound that includes an oxoanion of tungsten or is a mixed oxide containing tungsten. The simplest tungstate ion is WO_4^{2-} [133]. Ortho tungstates feature tetrahedral W(VI) centers with short W–O bond distances of 1.79 Å. Tungsten has six coordinate sites. It is a greyish-white lustrous metal and it is in solid form at room temperature. The melting point of tungsten is highest, at 3410 °C, and the vapor pressure lowest of all metals, and the highest tensile strength is found at 1650 °C. It has excellent corrosion resistance. The atomic number and weight of tungsten are 74 and 183.86, respectively. Therefore, the electron configuration of tungsten is $1s^2\, 2s^2\, 2p^6\, 3s^2\, 3p^6\, 4s^2\, 3d^{10}\, 4p^6\, 5s^2\, 4d^{10}\, 5p^6\, 6s^2\, 4f^{14}\, 5d^4$. It has a body-centered cubic structure. The importance of tungstates is based on their unique structural, electrical, thermal, and spectroscopic properties, which have potential applications in the field of optoelectronics devices and technology. They are promising hosts for luminescence studies due to their excellent properties, such as chemical and thermal stability [134]. Thus, among the different oxide hosts, tungstate and molybdate hosts are well known, and they are highly suitable for rare earth doping. When trivalent lanthanide ions are doped with this host, they can be used as luminophores and laser devices as well as in solar cells, not only as single crystals but also as powders. Thus, they show excellent optical properties doped with trivalent rare earth ions and are widely studied in diluted compounds. Besides the different hosts used in UC and DC mechanisms, tungstate and molybdate are an excellent matrix due to their tetragonal crystal system, better chemical stability, and low phonon energy [135, 136]. Recently, Xu et al. [137] have reported that $Ho^{3+}/Yb^{3+}/Tm^{3+}$ triple doped $CaWO_4$ polycrystalline phosphor shows white light emission, while William Barrera et al. [138] have demonstrated that rare earth (rare earth = $Ho^{3+}/Yb^{3+}/Tm^{3+}$)–doped nanocrystalline $KLu(WO_4)_2$ phosphor exhibits red, green, and blue emissions under NIR range (930 nm) laser excitation. However, Nd^{3+}-doped $KGd(WO_4)_2$ and $KY(WO_4)_2$ crystals are very prominent host matrices, which show NIR emission at 1067 nm [139–141]. Other than this, $KGd(WO_4)_2$: Ho^{3+} and $KYb(WO_4)_2$: Tm^{3+} are well-known laser crystals in which laser action has been demonstrated successfully [142, 143]. Thus, Kaminskii et al. reported that potassium lanthanide double tungstate and molybdate crystals are excellent luminescent hosts for both rare earth and transition metal ions [144, 145]. Interesting optical properties are shown in metal cation doped rare earth ions (MRE) $(XO_4)_2$ (M=Cs, K, Rb; X=Mo, W) phosphor when doped with rare earth ions in relation to phonon properties, energy transfer, and energy migration, including the site symmetry, in one-, two-, and three-dimensional systems [146–150]. Therefore, phosphors based on rare earth–doped tungstates play a vital role because of their tremendous performance in the fields of lasers, catalysis, and ionic conductors, and they are potential candidates due to the presence of the WO_4^{2-} group [151]. They have the ability to show a self-emission band in the blue-green region [152]. Ln^{3+}-activated tungstate phosphors could emit NIR as well as white light, which has more significance in the solar cell and solid-state lighting applications [153].

2.11.2 Vanadates

In chemistry, a vanadate is a stable host containing an oxoanion of vanadium. It has the highest oxidation state, that is, +5. The structure of the simplest vanadate ion is tetrahedral [154]. The structure and bonding in vanadate are represented by a single double bond; however, this is a resonance form, and it is a regular tetrahedron with four equivalent oxygen atoms attached. There exist other vanadates, such as $RhVO_4$, which has a statistical rutile structure in which the Rh^{3+} and V^{5+} ions randomly occupy the Ti^{4+} sites in the rutile lattice [155, 156]; it does not contain a lattice of cations and balance the vanadate anions; they are mixed oxides. From their chemical structure, it is seen that the central vanadium atom is surrounded by four anions. Thus, among the various host materials, the rare earth–activated alkaline earth metal vanadates show excellent luminescence properties in the visible to NIR spectral region. For example, a vanadate phosphor with the chemical formula $M_3(VO_4)_2$(M=Mg, Ca, Sr and Ba) is being considered as a good candidate for luminescent study because of its ability to strongly absorb UV light and then transfer the energy to other Ln^{3+} ions due to their well-matched energy levels in a wide region, and it shows a strong interaction between energy states of VO_4^{3-} and Ln^{3+} ions [157] in different host matrixes such as YVO_4, $Mg_3(VO_4)_2$, $LiZnVO_4$, $Na_2YMg_2(VO_4)_3$, and $NaCaVO_4$ [158–162]. Hence, the VO_4^{3-} group is surrounded by four O^{2-} ions in tetrahedral symmetry, similar to other Scheelite-type structures. Therefore, this type of vanadate-based host material shows efficient NIR emission when doped with Nd^{3+}, Pr^{3+}, Yb^{3+}, or Sm^{3+} ions, which is useful to increase the efficiency of c-silicon solar cells via the UP/DC process. A well-known wakefieldite-type (YVO_4) host provides favorable conditions as a luminescent host material due to efficient resonant energy transfer from host to lanthanide ions when it is doped with Yb^{3+} ion. Also, it has been noted that there exists an efficient cooperative energy transfer (CET) from YVO_4 host to Yb^{3+}, and hence, the intensity of Yb^{3+} emission was significantly improved in yttrium orthovanadate [163]. A new family of vanadates has the formula Ca_9 $R(VO_4)_7$ (R=Bi^{3+}) or a trivalent rare earth [164–166], which shows good chemical and physical properties and melts congruently. These compounds find potential application for non-linear laser crystals, and the $Ca_9Y(VO_4)_7$ host is one of them, which has an excellent integrated performance. In such a host, there is a statistical distribution of Y^{3+} and Ca^{2+} atoms, and it has a partial disordered structure. When Er^{3+} ions are doped into a host lattice, they occupy the positions of Y^{3+} ions, and hence, in this case, the broadening of the absorption and emission lines of Er^{3+} ions can be expected. Therefore, a new Er^{3+}-doped $Ca_9Y(VO_4)_7$ laser crystal is obtained.

2.11.3 Oxides

Research on oxide-based inorganic luminescence materials having high efficiency and good thermal and chemical stability has revealed wide interest over the last few years. The emission characteristics of rare earth ion–doped luminescent materials have been reported in the optical region of the electromagnetic spectrum, and they have been observed through UC and DC processes. Thus, we know that UC phosphors are very useful for photonics applications, such as display devices, solid-state lighting, solar cells, biological applications, and so on. Among the different oxide-based host materials, mesostructured TiO_2 or Al_2O_3 planar electrodes have been widely used in solar cells as hole blocking electrodes. ZnO is also known for its electron mobility, which is substantially higher than that of TiO_2 [167–169], which makes it an ideal material for an electron selective contact layer in solar cells. Dong et al. [170] reported that the efficiency for a perovskite solar cell reached 10.7% with the help of ZnO nanowires as an active layer, whereas Mahmood et al. [171] achieved an efficiency of 10.8 % via ZnO thin film doped with aluminum deposited by electro-spray as an electron selective layer. In addition to this, ZnO is of great interest in optoelectronic devices due to the fact that thin films of this material can be easily prepared and used for doping with various elements (Al, Co, Ga); also, the film can be deposited by different techniques, such as spray pyrolysis, sol-gel process, sputtering, chemical spray, plasma enhanced chemical vapor deposition (CVD), pulsed laser chemical spray, CVD and pulsed laser deposition, and so on [172].

2.11.4 Perovskites

Halide perovskites are a class of compounds with the general chemical formula ABX_3, where A represents an organic ammonium cation (e.g., CH_3NH_3), B is a metal (such as Pb or Sn), and X is a halide anion. Perovskite materials exhibit peculiar properties, so that they are very important and attractive in the fabrication of solar cells, and they show excellent optical properties, which are tuned by changing the composition of metals [173]; also, they are better charge transport materials and have long electron–hole diffusion lengths [174]. The synthesis of perovskite compounds is easy, and techniques for their deposition are also simple [175]; some well-known techniques are the one-step spin-coating method and the two-step sequential method [176], vapor deposition under high vacuum [177], and so on. Thus, solar cells based on halide perovskite compounds can show excellent properties in mesoscopic or planar heterojunction configurations [178–181]. A well-known methylammonium lead iodide ($CH_3NH_3PbI_3$) has a band gap of 1.5–1.6 eV. It has the capacity to strongly absorb light extending up to a wavelength of 800 nm, and it has potential application as a light harvester in solar cells [182]. Many researchers are working on perovskite cells and also trying to improve their efficiency. The efficiency of $MAPbI_3$ was improved by loading it on a mesoporous TiO_2 electrode by sequential deposition of PbI_2 and methylammonium iodide (MAI), when its efficiency reached up to 15% under 1 sun illumination [183]. The maximum efficiency in planar devices is reported to be 19.3% [184]. However, most perovskite-type materials suffer from degradation under moisture, oxygen, and ultraviolet radiation [185]. Therefore, $MAPbI_3$-based cells were found to be unstable under atmospheric conditions, and hence, their commercial uses are limited. Hence, there is an urgent demand for novel and thermally stable perovskite phases for solar cell applications, which will fulfill all the requirements for an ideal cell. Among these compounds, the double perovskite structured oxide compounds ($A_2BB'O_6$) have been widely studied in the past few years due to the diversity in their crystal structure and a variety of properties, such as high temperature superconductivity and colossal magneto-resistance, and they are of great importance for scientific and technological applications [186, 187]. These materials have become key materials in developing oxide-based electronic devices because of some important characteristics, such as high dielectric constant, ferroelectric polarization, catalysis, and magnetic media [188, 189]. Also, the research based on double perovskite–type compounds in the area of fuel cells is very important. These are promising anode materials for solid-state fuel cells [190, 191]. A newly emerging field in the area of solar cells is the hybrid inorganic-organic halide perovskite solar cells (PSCs), which show great success and are viewed as "the next big thing" in photovoltaics [192–197] in comparison with the Cu (InGa)Se$_2$[198–200], CdTe [201–204] thin film solar cells that have been used from the past few decades. In PSCs, the perovskite layer has been used as both the light absorber and the charge transporter [205–211].

2.11.5 Aluminates

Aluminates are an important class of luminescent materials used from the past few decades in fluorescent lamps and plasma display panel devices. Different types of aluminate systems are available for application in lighting technology; they have excellent thermal stability and better physical and chemical properties. Aluminate-based phosphor materials are further classified as binary alkaline earth aluminates such as $SrAl_2O_4$ [212, 213], hexaaluminate-based phosphors [214–222], and hexaaluminates with the magnetoplumbite structure [223–225]. The structure of MAl_2O_4 ($M = Ba^{2+}$, Sr^{2+}, Ca^{2+}) compound is formed of a three-dimensional framework of corner-sharing AlO_4 tetrahedrons, each oxygen being shared with the corresponding two aluminum ions, so that each tetrahedron has one net negative charge. The charge balance is accomplished by the divalent cations, and they occupy the interstitial sites within the host lattice. Among these aluminate hosts, a monoaluminate such as $SrAl_2O_4$ undergoes a phase transition from the monoclinic distorted structure to the hexagonal tridymite structure from low to high temperature [226].

A stuffed tridymite structure such as $CaAl_2O_4$ has two crystallographic nine-fold coordination sites for the large cations. The binary hexaaluminate compounds $MAl_{12}O_{19}$, $CaAl_{12}O_{19}$, and $SrAl_{12}O_{19}$ have the magnetoplumbite structure [227, 228]. In the case of a beta-alumina structure such as $BaAl_{12}O_{19}$ compound, this type of compound does not actually exist; instead, there are two compounds, one deficient in barium and one with excess barium [229, 230], and it has two sites: one is a large cation 12-coordinated in the magnetoplumbite structure and the other is 9-coordinated in the beta-alumina structure. Again, it is found that a set of more complicated superstructures is obtained when MgO is added as a third component. Therefore a set of more complicated superstructures appears that are based on stacking sequences of magnetoplumbite, beta-alumina, and spinel building block stacking sequences. In the last few decades, many compositions of the phosphor have been investigated by researchers as potential phosphor hosts, but the number of distinct compounds and their compositions were uncertain.

However, the aluminates are well-known luminescent display materials, which are evergreen among the family of hosts, because they can be synthesized by any methods, and products of many shapes and sizes can be produced. The compound with general formula $SrAl_2O_4$ has a series of advantages: it has stable physical and chemical properties [231], it shows long lifetime and high radiation intensity when doped with suitable rare earth ions [232], and its NIR emission is useful for application in c-Si solar cells. Rare earth (Eu^{2+}–Yb^{3+})–doped $SrAl_2O_4$ phosphors have been reported as efficient DC phosphors used in c-Si solar cells.

2.11.6 Borates

Borates are represented by a large number of boron-containing oxyanions. They are composed of trigonal planar BO_3 or tetrahedral BO_4 structure joined by surrounding oxygen atoms, which may be cyclic or linear in structure. On the basis of valence bond theory, the bonding in the structure is formed by sp^2 hybrid orbitals. Among other borate compounds, orthoborates do not contain the trigonal planar ion, but the high-temperature form contains planar BO_3^{3-} ions. Borate-based compounds have attracted considerable interest due to their outstanding physical [233–237], piezoelectric, and optical properties [238, 239]. Among MO–Bi_2O_3–B_2O_3 systems, where

M = alkaline earth metal, the $BaO-Bi_2O_3-B_2O_3$ ternary system is interesting due to the well-known nonlinear optical (NLO) compound b-BaB_2O_4 [240], the highly promising material BiB_3O_6 [241–243], and noncentrosymmetric compounds such as $BaBiBO_4$ [244]. Egorysheva and co-workers reported the phase equilibrium in the temperature range of 600–700 °C for four ternary compounds, $Ba_3BiB_3O_9$, $BaBiBO_4$, $BaBi_2B_4O_{10}$, and $BaBiB_{11}O_{19}$, in the $BaO-Bi_2O_3-B_2O_3$ system [245, 246]. Also, Bubnova et al. reported the crystal structure of $BaBi_2B_4O_{10}$ from single crystal X-ray diffraction data and carried out investigations on thermal expansion by high-temperature X-ray powder diffraction [247]. Today, there is a need for new optical functional materials synthesized at low temperature [248–251]. The $BaO-Bi_2O_3-B_2O_3$ system finds a special place among the resources of optical functional materials. Besides this, low-temperature modifications of b-BaB_2O_4 [252] and LiB_3O_5 [253], compounds are used for the harmonic generation of laser radiation in the visible and UV spectral range, and due to this advantage, interest in them is growing rapidly. Many hosts have been studied, and a series of new crystals with interesting optical properties has been produced, such as $CsLiB_6O_{10}$ (CLBO), $GdxY_{1-x}Ca_4O(BO_3)_3$ (GdYCOB), $Sr_2Be_2B_2O_7$ (SBBO), $K_2AlB_2O_7$ (KAB), and fluoride borates $KBe_2BO_3F_2$(KBBF) and $BaAlBO_3F_2$(BABF) [254–257]. These materials show excellent photoluminescence properties and are regarded as useful phosphor materials for many applications [258–261]. It is a challenging task to develop a new borate phosphor with promising luminescence and optical properties. Alkaline earth borates are important luminescent materials due to their excellent chemistry, thermal stabilization, and facile synthesis, and the raw material (H_3BO_3) is also cheap, so that they have been extensively used in solid-state lighting technology. To date, there are few reports available of borate-based phosphors efficiently excited by LED chips and applications of this material for white LEDs [262, 263]. Among these different characteristics, some work has been published on DC luminescence in rare earth–doped (Ce^{3+}, Yb^{3+}, Pr^{3+}, etc.) borate phosphors, showing that they are promising hosts to improve the efficiency of silicon solar cells [264].

2.11.7 Silicates

Rare earth ions (Er^{3+}, Yb^{3+}, Ce^{3+}, Ho^{3+}, Tb^{3+}, Eu^{3+}, Sm^{3+}, Dy^{3+}) have gained considerable attention during the past few decades for their superior optical properties when doped with host lattices. When rare earth ions are introduced into a suitable host, it shows an efficient emission with high quantum yield, narrow bandwidth, and large Stokes shifts, converting unusable UV to useful visible light. Among the different oxide-based host lattices, silicate-based phosphors have attracted much attention because of their many advantages, such as water resistance and color variety, high melting temperature, good thermal stability, physical and chemical properties, and so on. Today, there are different types of host available that have been studied as luminescence materials for solid-state lighting applications. From the different subgroups of silicate hosts, forsterite (Mg_2SiO_4) is an important class of material in the magnesia–silica system; it belongs to the olivine family of crystals, which has an orthorhombic crystalline structure, where Mg^{2+} occupies two non-equivalent octahedral sites in the crystal lattice, such as (M_1) with inversion symmetry (C_I) and the other (M_2) with mirror symmetry (CS). This type of material has some essential properties, such as high melting point, high chemical stability, wide electrical and refractory characteristics, good mechanical properties, bioactivity and biocompatibility, and so on [265]. It finds potential applications in many industrial areas, such as in electronics as an insulator, working at high frequencies [266], refractory industry [267], advanced technologies and so on [268], biomedicine [269], and luminescent technology [270]. Also, nanoparticles of rare earth ion–activated Mg_2SiO_4 show excellent photo-, thermal and sonoluminescence properties. From the reported literature, it is found that Mg_2SiO_4 doped with various rare earth ions finds a wide range of applications in the field of display devices and as energy harvesting materials for solar cell application [271–275]. In 2014, Sun et al. reported Eu^{2+}-doped barium silicate (Ba_2SiO_4: Eu) nanophosphors as downshifting materials for Si solar cells. The prepared nanophosphors are coated on the front side of the solar cell, and again it is coated with a metal-enhanced layer composed of Ag nanoparticles with a SiO_2 spacer. Thus, when this phosphor is applied to a Si-solar cell surface, the luminescence intensity is improved by the metal-plasmonic enhancement. The short-circuit current of the cell was increased to $0.86 \, mA/cm^2$, and the powder conversion efficiency reached 0.64%. However, Cattaruzza et al. introduced silicate glass slides doped with silver or copper by ion exchange in molten salts and showed their use as a cover glass for a GaAs-based solar cell. This rare earth ion–doped glass exhibited downshifting properties by absorbing UV and near UV light and emitting broad band visible light. When these silver- and copper-doped glasses were separately coated on the solar cell surface, the copper-doped glass showed higher output power than the silver-doped glass. The efficiency of the cell reached 2%. Hence, it is indicated that silicate-based materials are also promising candidates and used in many display and energy applications.

2.11.8 Aluminosilicates

Alumina (Al_2O_3) and silica (SiO_2) are the two most plentiful minerals in the earth's crust. The class of minerals containing aluminum oxide and silicon oxide is called *aluminosilicates*. Their chemical formulae are often expressed as $xAl_2O_3.ySiO_2.zH_2O$. The study of aluminosilicate-based phosphors started in 1988. Anorthite ($CaAl_2Si_2O_8$), having a triclinic crystal system with space group I_1 [276], was the first aluminosilicate phosphor, reported by Angel et al. in 1988. Later, some new calcium aluminosilicate materials were reported, based on the combination of $CaO–Al_2O_3–SiO_2$ ternary system, and many of these were selected as the host resources for efficient phosphors, such as

$CaAl_2Si_2O_8:Eu^{2+}$ [277], $CaAl_2Si_2O_8:Eu^{2+}$, Mn^{2+}[278], $CaAl_2Si_2O8:Eu^{2+}$, Ce^{3+} [279], and $Ca_2Al_2SiO_7:Ce^{3+}$, Tb^{3+} [280]. There are different types of aluminosilicates based on their structure, such as andalusite, kyanite, and sillimanite, having the chemical formula Al_2SiO_5 $(Al_2O_3.SiO_2)$. The name kyanite is derived from the Greek word *kuanos*, at times referred to as *kyanos*, meaning deep blue, and sillimanite is named after the American chemist Benjamin Silliman (1779–1864) [281]. These have different crystal structures. Another group of aluminosilicates is feldspar, in which the Si^{4+} ions in silicates are replaced by Al^{3+} ions; hence, the charge must be balanced by other positive ions such as Na^+, K^+, and Ca^{2+} ions. Some examples of the feldspar group are

- Sanidine $[(K, Na)AlSi_3O_8]_4$
- Orthoclase $[(K, Na)AlSi_3O_8]_4$
- Albite $[NaAlSi_3O_8]_4$
- Anorthite $Ca[Al_2Si_2O_8]$

In this group, the Al^{3+} ions replace the Si^{4+} ions in the chains of corner shared tetrahedrons. However, the bonding between Al and Si can be different. Silicon atoms tend to bond by four oxygen atoms in a tetrahedral fashion, but aluminum ions tend to be bonded to six oxygen atoms in an octahedral fashion. The compound $Al_2Si_2O_5$ $(OH)_4$, $(Al_2O_3.2SiO_2.2H_2O)$ occurs naturally as the mineral kaolinite. It is also called aluminum silicate dehydrate [282]. It is a fine white powder and is used as a filler in paper and rubber and also in paints [283]. The other compound, $Al_2Si_2O_7$, $(Al_2O_3.2SiO_2)$, called metakaolinite, is formed from kaolin by heating at 450 °C (842 °F) [284]. Al_6SiO_{13}, $(3Al_2O_3.2SiO_2)$ is the mineral mullite, having only a thermodynamically stable intermediate phase in the Al_2O_3–SiO_2 system at atmospheric pressure [285]. This is also called 3:2 mullite to distinguish it from $2Al_2O_3.SiO_2$, Al_4SiO_8 (2:1 mullite).

The few important characteristics of aluminosilicate-based phosphor are

- Unique spectroscopic behavior
- High chemical stability
- High thermal stability
- Water resistant
- Weather resistant
- High absorption in UV region

In recent years, alkaline earth aluminosilicates have been widely studied and have received great attention in the field of luminescent materials. Aluminate-based luminescent materials are not more chemically stable; and their properties are affected to a great extent when they come into contact with moisture for several hours, and hence, there are some limitations for actual applications [286]. The chemical and thermal stability of a phosphor is important for knowing the degradation properties of materials at high temperature, which affect the quality and efficiency of the device [287]. Aluminosilicate-based luminescent materials have noteworthy advantages over other phosphors on account of their physical and chemical stability, wide-ranging luminescent colors, and exceptional water-resistant properties [288]. Excellent luminescent performance for PDPs and FEDs has been reported by Kubota et al. for $Sr_3Al_{10}SiO_{20}:Eu^{2+}$ phosphor [289]. An $NaAlSiO_4:Ce^{3+}$, Mn^{2+} aluminosilicate-based phosphor has been reported recently by Junhezhou et al. with excellent properties. $Ca_2Mg_{0.5}AlSi_{1.5}O_7:Ce^{3+}$, Eu^{2+}, $Ca_3Al_2Si_2O_8Cl_4$: Eu^{2+}, Mn^{2+}, $Ca_2Al_2SiO_7:Ce^{3+}$, and Tb^{3+} are some aluminosilicate-based phosphors reported in the literature.

2.11.9 Fluorides

Recently, Ln^{3+}-doped UC glass-ceramics have received significant interest due to their potential application in spectral conversion, temperature sensors, and solar cells [290]. Thus, similarly to the above-discussed hosts, transparent oxyfluoride glass-ceramics are important candidates due to their extensive optical properties and mechanical stability [291]. The optical properties of the Ln^{3+} doping ions will be controlled by the fluoride host materials. To date, different types of fluoride hosts have been studied as luminescent materials for application in lighting and thermoluminescence dosimetry. Among these, $NaYF_4$ LaF_3 and CaF_2 are well-known fluoride nanocrystals, which show excellent UC properties when doped with Er^{3+}, Yb^{2+}, or Nd^{3+} ions. Also, they can show a high-intensity pulse laser beam when doped with certain other impurity ions. CaF_2 is well known due to its desirable properties, such as low refractive index (RI), low phonon energy, high laser damage threshold, and so on [292]. So, these materials have more multifunctional properties than other host materials used in the past few decades.

2.11.10 Nitrides

Phosphor-converted white LEDs are the most promising and famous sources of white light. They are fabricated by the combination of LEDs coupled with inorganic phosphors. Here, phosphor is the main component, which plays a key role in the overall performance, such as QEs, CRI, color temperature, and durability of the devices [293]. Currently, white LEDs fabricated using yellow (YAG:Ce^{3+}) phosphor suffer from drawbacks due to the presence of red color in the emission band of YAG:Ce^{3+}, so that it is not suitable for use in lighting devices. Therefore, new phosphors are always in demand to upgrade the device output. Nowadays, there are some hosts that are practically used in SSL technology, such as oxides, sulfides, fluorides, and nitrides [294]. Among hosts of this kind, nitride-based phosphors are receiving great attention due to their excellent optical properties, high quantum efficiency, good

CRI, and better stability at high temperature. Thus, many scientists are busy finding novel nitride phosphors for LEDs. Some of the reported and commercially used nitride phosphors are $AlN:Eu^{2+}$, β-sialon:Eu^{2+}, α-sialon:Eu^{2+}, $CaAlSiN_3:Eu^{2+}$, $La_3Si_6N11:Ce^{3+}$, and $Sr_2Si_5N_8:Eu^{2+}$ [295]. At present, the discovery of novel nitride phosphors with sharp narrowband emission is preferred, so that in future, they will have significant advantages in improving the performance of display devices. Thus, similarly to the nitride phosphors, oxynitride phosphors are also interesting and more demanding hosts due to the need for efficient color-conversion materials in LEDs applications [296]. In the past few years $SrSi_2O_2N_2:Eu$ is a well-known green conversion phosphor material used in optical devices [297]. Hence, it is also an excellent host for white LEDs. However, although the oxynitride phosphors have excellent chemical stability, their synthesis methods are complex, which is the main disadvantage associated with this system.

2.12 Other Materials for LED Applications

2.12.1 Organic LEDs

Organic materials consist of covalent molecules; covalent bonds exist between their atoms, so that the intermolecular and interatomic van der Waals and London forces are weaker. Thus, the charge transport bands in organic materials are narrow as compared with inorganic compounds, and the band structure is easily broken by adding defects in the molecular system. Organic materials consist of π-conjugated systems, which are formed by the overlap of P_z orbitals. Therefore, the electrons are delocalized within a molecule, the energy gap between the highest occupied orbital (HOMO) and lowest unoccupied orbital (LUMO) is relatively small, and emission is observed in the visible range. Generally, the materials used in the study of organic electroluminescence devices are small organic molecules or polymers. OLEDs are fabricated using polymers or small organic molecules, because they are more energy efficiently active elements. Recently, organic devices have been used in many display systems, and they can be fabricated in any form due to their flexible substrates. Organic LEDs are classified into two classes: OLEDs (use of small organic molecule) and PLEDs (organic polymer). The typical double heterostructure OLED, which is fabricated from small molecules, consists of three organic layers sandwiched between the cathode and the anode. The electron transport layer (ETL) and hole transport layer (HTL) are placed adjacent to the cathode and the anode, respectively. The emissive layer (EML) plays an important role in light emission. Generally, it consists of emissive dyes or impurity ions doped into a suitable host material (just like HTL or ETL material) [298, 299].

2.12.2 Polymer LEDs

Current lighting technologies based on organic semiconductors are increasingly demand, and they cover a large market in Asian and other developing countries in the field of semiconductor display devices due to their special advantages, such as lightweight displays, portable computing, and so on. These types of device have cheap fabrication techniques, and the device efficiency is also better. Hence, considering these benefits of polymer materials, it is seen that PLEDs will be a better substitute in the near future for the development of energy-saving and friendly lighting technology. PLEDs have a relatively simple fabrication process; they are formed by the combination of a light-emitting polymer (LEP) layer, host materials, and a hole injection layer (HIL), and deposited on an Indium Tin Oxide (ITO)–glass substrate. Here, poly (phenylene vinylene) (PPV) and polyfluorene (PFO) polymers are used as an LEP layer to fabricate the device. Therefore, a PLED is an dual-carrier injection device in which the electrons are injected from the cathode to the LUMO orbital of the polymer and holes are injected from the anode to the HOMO orbital of the conducting polymer; the electrons and holes recombine radiatively within the polymer, producing light. To upgrade the device performance of PLEDs, researchers are making efforts to find different kinds of electrodes and are adding more layers (such as nanoparticle layers) between the electrodes [300, 301].

2.12.3 Quantum Dots for LEDs

Inorganic nanostructure materials are of great interest due to their applications in many cutting-edge technologies [302]. Therefore among the different nanostructures used in semiconductor devices, nanoscale phosphors are also interesting materials, and they are widely used in solid-state lasers, LEDs, and planar display devices [303]. The phosphors used in the display have some special characteristics, such as high contrast, better resolution, thermal stability, packing density, and adhesion, which can be fulfilled only by the use of nanosized phosphors [304]. Therefore, with further developments in the device technology, the demand for higher-performance, lightweight, and low–power consumption display devices is increasing worldwide. Recently, colloidal quantum dots (QDs) have become promising candidates for next-generation optoelectronic materials due to their extraordinary properties, such as tunable emission, narrow bandwidths, high brightness and efficiency, and so on. Thus, more efforts have been made by researchers to develop high-quality QDs and QD-based photoelectric devices such as LEDs [305], display panels [306], solar cells [307], and lasers [308]. Therefore, among the other display devices, OLEDs and QD-LEDs are promising sources that have the potential to fulfill the criteria of new emerging display technologies. QD-LEDs have better optical properties than OLEDs. QLEDs show intense white light emission, and hence, they have attracted much attention in the last 10 years. If we compare QLEDs with traditional phosphor-based white LEDs, it is found that QLEDs have more advantages regarding some drawbacks of phosphor-based LEDs, such as low CRI and high energy consumption [309]. Hence, during the past few decades, research has focused on the development of new QDs composed of binary compounds [310]. But the challenge still remains to maintain the optical properties of binary QDs during device fabrication to avoid surface traps, which are formed due to long-term operation, and nonradiative relaxations during emission [311]. CdZnS QDs are ideal due to band edge absorption, high quantum yields, narrow spectral band, and high stability. The properties of QLEDs based on CdZnS/ZnSe QDs show excellent results compared with other reported QDs. Therefore, CdZnS/ZnSe QDs may be a promising candidate for the future development of white QLEDs.

References

1. C. G. A. Hill, CRT Phosphors, SID Seminar Lecture Notes 2 (1984) S-6.
2. H. S. Nalwa, L. S. Rohwer, *Handbook of Luminescence, Display Materials, and Devices,* Vol. 3: Display Devices, California: American Scientific Publishers (2003).
3. E. Harvey, Newton (Edmund Newton), 1887–1959 A history of luminescence from the earliest times until 1900.
4. S. Shionoya, W. Yen, *Phosphor Handbook,* Boca Raton, FL: CRC Press (1999).
5. M. H. V Werts, *Sci. Progr.* 88 (2005) 101.
6. J. Rumble, Handbook of Chemistry and Physics, 99th Edition (2018). ISBN 9781138561632) Boca Raton, FL: CRC Press 36.
7. K. N. Shinde, S. J. Dhoble, H. C. Swart, K. Park, *Phosphate Phosphors for Solid-State Lighting* (DOI: 10.1007/978-3-642-34312-4_2), 41 Springer 174.
8. B. Geijer, *Ann. fur die Freunde der Naturlehre* 9 (1788) 229.
9. W. H. Brock, *The Norton History of Chemistry,* 1st edn. W. W. Norton & Company (1993).
10. R. T. Wegh, A. Meijerink, R. J. L. Kki, J. H. Klsa, *J. Lumin.* 87–89 (2000) 1002.
11. R. Visser, P. Dorenbos, C. W. E. Van Eijk, A. Meijerink, H. W. den Hartog, *J. Phys. Condens. Matter* 5 (1993) 8437.
12. S. Kubodera, M. Kitahara, J. Kawanaka, W. Sasaki, K. Kurosava, *Appl. Phys. Lett.* 69 (1996) 452.
13. G. Blasse, B. C. Grabinaier, *Luminescent Materials.* Berlin: Springer-Verlag, Chapter 6 (1996).
14. R. T. Wegh, H. Donker, A. Meijerink, R. J. Lamminmaki, J. Holsa, *Phys. Rev. B* 56 (1997) 13841.
15. E. Loh, *Phys. Rev.* 175 (1968) 533.
16. C. Dujardin, B. Moine, C. Pedrini, *J. Lumin.* 54 (1993) 259.
17. H. Yu, Y. W. Lai, G. M. Gao, L. Kong, G. H. Li, S. C. Gan, G. Y. Hong, *J. Alloys Compd.* 509 (2011) 6639.

18. G. Blasse, B. C. Grabmaier, *Luminescent Mater.* (ISBN 978-3-642-79017-1) 46 (1994).
19. W. J. Dong, Y. J. Zhu, H. D. Huang, L. S. Jiang, H. J. Zhu, C. R. Li, B. Y. Chen, Z. Shi, G. Wang, *J. Mater. Chem. A.* 1 (2013) 10030.
20. M. M. Shang, G. G. Li, D. L. Geng, D. M. Yang, X. J. Kang, Y. Zhang, H. Z. Lian, J. Lin, *J. Phys. Chem. C.* 116 (2012) 10231.
21. Q. Li, V. W. W. Yam, *Angew. Chem. Int. Ed.* 46 (2007) 3486.
22. M. Nogami, T. Nagakura, T. Hayakawa, *J. Lumin.* 86 (2000) 117.
23. X. Fan, M. Wang, *Mater. Sci. Eng. B,* 21 (1993) 55.
24. R. Balda, J. Fernandez, A. J. Gracia, G. F. Imbusch, *J. Lumin.* 76 and 77 (1998) 551.
25. S. Tanabe, K. Hirao, N. Soga, *J. Non Cryst. Solids* 113 (1989) 178.
26. B. R Judd, *Phys. Rev.* 127 (1962) 750.
27. Ofelt G. S. J., *Chem. Phys.* 37 (1962) 511.
28. G. H Dieke, edited by Hannah Crosswhite. *Spectra and Energy Levels of Rare Earth Ions in Crystal.* Wiley (1968).
29. C. Feldmann, T. JuEstel, C. R. Ronda, P. J. Schmidt, *Adv. Funct. Mater.* 13 (7) (2003) 511.
30. G. Blasse, B. C. Grabmaier, *Luminescent Materials.* Berlin: Springer-Verlag (1994) 40.
31. W. Jingjing, S. Shi, X. Wang, J. Li, R. Zong, W. Chen, *J. Mater. Chem. C* 2 (2014) 2786.
32. B. R. Judd, *Phys. Rev.* 127 (1962) 75.
33. G. S. Ofelt, *J. Chem. Phys.* 37 (1962) 511.
34. M. F. Gui, Z. X. Min, S. H. Jin, *Opt. Laser Technol.* 44 (2012) 185.
35. H. M. Gyu, B. M. Rang, H. T. Eun, B. J. Seong, K. Yangsoo, S. Park, Y. H. Soon, K. S. Hong, *Ceram. Int.* 38 (2012) 1365.
36. K. Li, X. Liu, Y. Zhang, X. Li, H. Lian, J. Lin, *Inorg. Chem.* 54 (2015) 323.
37. Z. Xia, D. Chen, *J. Am. Ceram. Soc.* 93 (2010) 1397.
38. Y. Yu, F. Song, C. Ming, J. Liu, W. Li, Y. Liu, H. Zhao, *Opt. Commun.* 303 (2013) 62.
39. G. H. Mhlongo, O. M. Ntwaeaborwa, H. C. Swart, R. E. Kroon, P. Solarz, W. R. Romanowski, K. T. Hillie, *J. Phys. Chem. C* 115 (2011) 17625.
40. A. Srivastava, A. Setlur, H. Comanzo, W. Beers, U. Happek, P. Schmidt, *Opt. Mater.* 33 (2011) 292.
41. A. Mayolet, W. Zhang, E. Simoni, J. C. Krupa, P. Martin, *Appl. Phys. B,* 4 (1995) 757.
42. J. E. Geusic, H. M. Marcos, L. G. Van Uitert, *Appl. Phys. Lett.* 4 (1964) 182.
43. V. Lupei, A. Lupei, N. Pavel, T. Taira, I. Shoji, A. Ikesue, *Appl. Phys. Lett.* 79 (2001) 590.
44. J. Lu, K. Ueda, H. Yagi, T. Yanagitani, Y. Akiyama, A. A. Kaminskii, *J. Alloys Compd.* 341 (2002) 220.
45. I. Oprea, H. Hesse, K. Betzler, *Phys. Status Solidi (b)* 242 (2005) R109.
46. W. X. Que, Z. Su, X. Hu, *J. Appl. Phys.* 98 (2005) 093518.
47. G. H. Dieke, *Spectra and Energy Levels of Rare-Earth Ions in Crystals.* New York, NY: Wiley (1969).
48. A. Miguel, R. Morea, M. A. Arriandiaga, M. Hernandez, F. J. Ferrer, C. Domingo, J. M. Fernandez-Navarro, J. Gonzalo, J. Fernandez, R. Balda, *J. Eur. Ceram. Soc.* 34 (2014) 3959.
49. R. Balda, A. Oleaga, J. Fernandaz, J. M. Fdez-Navarro, *Opt. Mater.* 24 (2003) 83.
50. S. G. Xiao, X. L. Yang, *Mod. Phys. Lett. B* 25 (4) (2011) 265.
51. H. H. T. Vu, T. S. Atabaev, N. D. Nguyen, Y. H. Hwang, H. K. Kim, *J. Sol-Gel Sci. Technol.* 71 (2014) 391.

52. T. S. Atabaev, Z. Piao, Y. H. Hwang, H. K. Kim, N. H. Hong, *J. Alloys Compd.* 572 (2013) 113.
53. D. Dosev, I. M. Kennedy, M. Godlewski, I. Gryczynski, K. Tomsia, E. M. Goldys, *Appl. Phys. Lett.* 1 (2006) 88.
54. P. A. Tanner, X. Zhou, F. Liu, *J. Phys. Chem. A* 108 (52) (2004) 11521.
55. X. P. Fan, D. B. Pi, F. Wang, J. R. Qiu, M. Q. Wang, *IEEE Trans. Nanotechnol.* 5 (2006) 123.
56. R. Nikifor, B. G. Renato, S. M. Glauco, *Mater. Res. Bull.* 74 (2016) 103.
57. M. Misiak, A. Bednarkiewicz, W. Strek, *J. Lumin.* 169 (2016) 717.
58. P. A. Loiko, N. M. Khaidukov, J. Mendez-Ramos, E. V. Vilejshikova, N. A. Skoptsov, K. V. Yumashev, *J. Lumin.* 170 (2016) 1.
59. M. H. Imanieh, I. R. Martín, A. Nadarajah, J. G. Lawrence, V. Lavín, J. Gonzalez Platas, *J. Lumin.* 172 (2016) 201.
60. S. K. Hussain, J. S. Yu, *J. Lumin.* 175 (2016) 100.
61. Q. Wang, J. Qiu, Z. Song, Z. Yang, Z. Yin, D. Zhou, J. S. Wang, *Am. Ceram. Soc.* 99 (2016) 911.
62. M. Ferhi, N. BenHassen, C. Bouzidi, K. Horchani-Naifer, M. Ferid, *J. Lumin.* 170 (2016) 174.
63. V. D. D. Cacho, L. R. P. Kassab, S. L. de Oliveira, N. I. Morimoto, *Mater. Res.* 9 (2006) 21.
64. Z. Antic, V. Lojpur, M. G. Nikolic, V. C. D. S. D. Signevic, P. S. Ahrenkiel, M. D. Dramićanin, *J. Lumin.* 145 (2014) 466.
65. Z. D. Fleischman, L. D. Merkle, G. Alex Newburgh, M. Dubinskii, *Opt. Mater.* 3 (2013) 1176.
66. M. Li, Y. Guo, G. Bai, Y. Tian, L. L. Hu, J. J. Zhang, *J. Quant. Spectrosc. Rad. Transfer.* 127 (2013) 70.
67. X. Qiao, H. J. Seo, *Mater. Lett.* 105 (2013) 166.
68. Y. Y. Zhang, J. Q. Deng, S. C. Ni, *Bull. Mater. Sci.* 36 (2013) 513.
69. D. Lande, S. S. Orlov, A. Akella, L. Hesselink, R. R. Neurgaonkar, *Opt. Lett.* 22 (1997) 1722.
70. A. V. Kir'yanov, V. P. Minkovich, Y. O. Barmenkov, M. A. Martinez Gamez, A. Martinez-Rios, *J. Lumin.* 111 (2005) 1.
71. S. Schietinger, T. Aichele, H. Wang, T. Nann, O. Benson, *Nano Lett.* 10 (2010) 134.
72. E. He, H. Zheng, J. Dong, W. Gao, Q. Han, J. Li, L. Hui, Y. Lu, H. Tian, *Nanotechnology* 25 (2014) 045603.
73. J. Li, J. Zhang, Z. Hao, X. Zhang, J. Zhao, Y. Luo, *J. Appl. Phys.* 113 (2013) 223507.
74. Q. Luu, A. Hor, J. Fisher, R. B. Anderson, S. Liu, T. S. Luk, H. P. Paudel, M. Farrokh Baroughi, P. S. May, S. Smith, *J. Phys. Chem. C* 118 (2014) 3251.
75. V. Mahalingam, R. Naccache, F. Vetrone, J. A. Capobianco, *Chem. Commun.* 47 (2011) 3481.
76. J. Nakanishi, T. Yamada, Y. Fujimoto, O. Ishii, M. Yamazaki, *Electron. Lett.* 46 (2010) 1285.
77. W. J. Chung, J. R. Yoo, Y. S. Kim, J. Heo, *J. Am. Chem. Soc.* 80 (1997) 1485.
78. B. G. Aitken, M. J. Dejneka, M. L. Powley, *J. Non-Cryst. Solids* 349 (2004) 115.
79. A. A. Kaminskii, *Crystalline Lasers: Physical Processes and Operating Schemes*, Boca Raton, FL: CRC Press (1996).
80. Y. Yang, B. Yao, B. Chen, C. Wang, G. Ren, X. Wang, *Opt. Mater.* 29 (2007) 1159.
81. D. L. Dexter, *J. Chem. Phys.* 21 (1953) 836.
82. D. L. Dexter, J. H. Schulman, *J. Chem. Phys.* 22 (1954) 1063.

83. S. Ye, F. Xiao, Y. X. Pan, Y. Y. Ma, Q. Y. Zhang, *Mater. Sci. Eng. R* 71 (2010) 1.

84. G. K. Liu, B. Jacquier, *Spectroscopic Properties of Rare Earths in Optical Materials.* Springer (2005).

85. H. S. Nalwa, L. S. Rohwer, *Handbook of Luminescence, Display Materials, and Devices.* California: American Scientific Publishers, 2 (2003).

86. W. M. Y. Shigeo Shionoya, *Phosphor Handbook.* Boca Raton, FL: CRC Press (1999).

87. Y. Q. Li, A. Delsing, G. de With, H. T. Hintzen, *Chem. Mater.* 17 (2005) 3242.

88. G. K. Liu, B. Jacquier, *Spectroscopic Properties of Rare Earths in Optical Materials.* Springer (2005).

89. A. Boukerika, L. Guerbous, *J. Lumin.* 145 (2014) 148.

90. R. Callister, Jr., *Materials Science and Engineering – An Introduction*, 8th edn. Wiley (2009) 5.

91. R. Callister, Jr., *Materials Science and Engineering – An Introduction*, 8th edn. Wiley (2009) 10.

92. C. E. Mortimer, *Chemistry: A Conceptual Approach*, 3rd edn. New York, NY: D. Van Nostrand Company (1975).

93. F. J. Duarte, *Appl. Opt.* 38 (1999) 6347.

94. M. A. McEvoy, N. Correll. *Science* 347 (6228) 2015.

95. C. Buzea, I. Pacheco, K. Robbie, Nanomaterials and nanoparticles: Sources and toxicity. *Biointerphases* 2(4) (2007) MR17.

96. S. Deng, Z. Xue, Q. Yang, Y. Liu, B. Lei, Y. Xiao, M. Zheng, *Appl. Surf. Sci.* 282 (2013) 351.

97. C. Wenisch, H. D. Kurland, J. Grabow, F. A. Müller, *J. Am. Ceram. Soc.* 99 (2016) 2561.

98. Z. L. Wang, *Mater. Sci. Eng. R* 64 (2009) 33.

99. V. C. Tung, L. M. Chen, M. J. Allen, J. K. Wassei, K. Nelson, R. B. Kaner, Y. Yang, *Nano Lett.* 9 (2009) 1949.

100. C. Y. Su, A. Y. Lu, Y. Xu, F. R. Chen, A. N. Khlobystov, L. J. Li, *ACS Nano*, 5 (2011) 2332.

101. Z. Liu, D. He, Y. Wang, H. Wu, J. Wang, *J. Synth. Met.* 160 (2010) 1036.

102. U. Caldiño, A. Speghini, S. Berneschi, M. Bettinelli, M. Brenci, E. Pasquini, S. Pelli, G. C. Righini, *J. Lumin.* 147 (2014) 336.

103. S. M. Lima, L. H. C. Andrade, A. C. P. Rocha, J. R. Silva, A. M. Farias, A. N. Medina, M. L. Baesso, L. A. O. Nunes, Y. Guyot, G. Boulon, *J. Lumin.* 143 (2013) 600.

104. J. Deng, J. Chen, Y. Liu, *J. Lumin.* 170 (2016) 835.

105. S. Zeng, D. Baillargeat, H. P. Ho, Yong. *Chem. Soc. Rev.* 43 (2014) 3426.

106. C. Stephenson, A. Hubler, Stability and conductivity of self assembled wires in a transverse electric field. *Sci. Rep.* (2015) 5, 15044.

107. A. Hubler, D. Lyon, Gap size dependence of the dielectric strength in nano vacuum gaps. *IEEE*, 20 (2013) 1467.

108. E. Pavitra, G. Seeta Rama Raju, J. H. Oh, J. S. Yu, *New J. Chem.* 38 (2014) 3413.

109. F. Li, L. Li, C. Guo, T. Li, H. Mi Noh, J. H. Jeong, *Ceram. Int.* 40 (2014) 7363.

110. V. Lojpur, P. Ahrenkiel, M. Dramicanin, *Nanoscale Res. Lett.* 8 (2013) 131.

111. R. Dey, V. K. Rai, *Dalton Trans.* 43 (2014) 111.

112. J. C. Boyer, F. Vetrone, L. A. Cuccia, J. A. Capobianco, *J. Am. Chem. Soc.* 128 (2006) 7444.

113. A. Kumari, V. K. Rai, K. Kumar, *Spectrochim. Acta Part A* 127 (2014) 98.

114. W. Elenbaas, *Philips. Tech. Rev.* 18 (1957) 167 (a); E. W. Philips Technical Library (1959).

115. M Tong, Y. Liang, P. Yan, Q. Wang, G. Li, *Opt. Laser Technol.* 75 (2015) 221.

116. R. Kane, H. Sell, *Revolution in Lamps: A Chronicle of 50 Years of Progress*, 2nd edn. Georgia: The Fairmont Press (2001) 93.

117. A. S. Brown, J. A. Henije, M. J. Delaney, *Appl. Phys. Lett.* 5 (1988) 1142.

118. S. Nakamura, G. Fasol, *The Blue Laser Diode*. Berlin: Springer (1997) (b) S. Nakamura, G., MRS Bulletin, 29 (1997) 16.

119 Y. G. Park, Luminescence and temperature dependency of β-SiAlON phosphor. Samsung Electro Mechanics Co.

120. H. Kume, Nikkei Electronics Sharp to Employ White LED Using Sialon. Nikkei Business Publications Sept 15 (2009).

121 H. Naoto, New sialon phosphors and white LEDs, *Oyo Butsuri* 74 (2005) 1449.

122. D. E. C. Corbridge, *The Structural Chemistry of Phosphorus*. Amsterdam: Elsevier (1974).

123. W. P. Hong, J. Singh, P. Bhattacharya, *IEEE Trans.* ED-7 (1986) 480.

124. Cathode ray tube [online]. Available from http://en.wikipedia.org/wiki/Cathode_ray_tube

125. The Cathode Ray Tube [online]. Available from http://electronics.Howstuffworks.com/tv3

126. H. A. Leverenz, F. Seitz, *J. Appl. Phys.* 10 (1939) 479.

127. H. A. Leverenz, *RCA Rev.* 5 (1940) 131.

128. L. K. Albert, F. C. Palilla, *Appl. Phys. Lett.* 5 (1964) 118.

129. R. A. Fields, M. Birnbaum, C. L. Fincher, *Appl. Phys. Lett.*, 51 (1987) 1885.

130. Flatpanel display [online]. Available from http://en.wikipedia.org/wiki/Flatpanel display

131. K. D. Oskam, R. T. Wegh, H. Donker, E. V. D. van Loef, A. Meijerink, *J. Alloys Compd.* 300 (2000) 421.

132. T. Trupke, M. A. Green, P. Würfel, *J. Appl. Phys.* 92 (2002) 1668.

133. E. Wiberg, A. F. Holleman *Inorganic Chemistry*. Elsevier. ISBN 0-12-352651 (2001).

134. A. M. Kaczmarek, K. V. Hecke, R. V. Deun, *Inorg. Chem.* 53 (2014) 9498.

135. W. Liu, J. Sun, X. Li, *Opt. Mater.* 35 (2013) 1487.

136. L. Lei, Y.Kuisheng, Z. Xiyan, Q.Ning, L. Hui, Z. Zhou, *J. Rare Earths* 30 (11) (2012) 1092.

137. Y. Xu, Y. Wang, L. Shi, *Opt. Laser Technol.* 54 (2013) 50.

138. E. W. Barrera, M. C. Pujol, C. Concepción, *Phys. Status Solidi* 9 (2011) 2676.

139. G. Boulon, G. Metrat, N. Muhlstein, A. Brenier, M. R. Kokta, L. Kravchik, Y. Kalisky, *Opt. Mater.* 24 (2003) 377.

140. A. Brenier, F. Bourgeois, G. Metrat, N. Muhlstein, G. Boulon, *Opt. Mater.* 16 (2001) 207.

141. G. Merat, N. Muhlstein, A. Brenier, G. Boulon, *Opt. Mater.* 8 (1997) 75.

142. O. Silvestre, M. C. Pujol, F. Gueell, M. Aguilo, F. Diaz, A. Brenier, G. Boulon, *Appl. Phys. B: Lasers Opt.* 87 (1) (2007) 111.

143. M. C. Pujol, F. Guell, X. Mateos, J. Gavalda, R. Sole, J. Massons, M. Aguilo, F. Diaz, G. Boulon, A. Brenier, *Phys. Rev. B* 66 (14) (2002) 144304.

144. A. A. Kaminski, P. V. Klevstov, L. Li, A. A. Pavlyuk, *Phys. Status Solidi* A5 (1971) K79.

145. A. A. Kaminskii, *Laser Crystals. Their Physics and Properties*. Berlin: Springer (1981).

146. M. Buijs, G. Blasse, L. H. Brixner, *Phys. Rev. B* 34 (1986) 12.

147. J. P. M. van Vilet, D. van des Voot, G. Blasse, *J. Lumin.* 42 (1989) 305.

148. P. A. M. Berdowski, G. Blasse, *J. Lumin.* 29 (1984) 243.

149. L. Macalik, *Pol. J. Chem.* 69 (1995) 286.

150. J. Hanuza, L. Macalik, B. Macalik, *Acta Phys. Pol. A* 84 (1993) 895, 899.

151. [a] L. Macalik, B. Macalik, W. Strek, J. Legendziewicz, *Eur. J. Solid Inorg. Chem.* 33 (1996) 397; [b] L. Macalik, J. Hanuza, J. Sokolnicki, J. Legendziewicz, *Spectrochim. Acta, Part A* 55 (1999) 251; [c] L. Macalik, J. Hanuza, J. Legendziewicz, *Acta Phys. Pol. A* 84 (1993) 909.

152. Y. Tian, B. Chen, R. Hua, N. Yu, B. Liu, J. Sun, L. Cheng, et al., *CrystEngComm.* 14 (2012) 1760.

153. A. M. Kaczmarek, K. V. Hecke, R. V. Deun, *Inorg. Chem.* 53 (2014) 9498.

154. C. F. Albert, W. Geoffrey, C. A. Murillo, M. Bochmann, *Advanced Inorganic Chemistry*, 6th edn. New York, NY: Wiley-Interscience (1999) (ISBN 0-471-19957-5).

155. N. N Greenwood, A. Earnshaw, *Chemistry of the Elements*, 2nd edn. Butterworth-Heinemann (1997) (ISBN 0-08-037941-9).

156. A. F. Wells, *Structural Inorganic Chemistry*, 5th edn. Oxford Science Publications (1984) (ISBN 0-19-855370-6) (1984).

157. H. P. Zhang, M. K. Lü, Z. S. Yang, Z. L. Xiu, G. J. Zhou, S. F. Wang, Y. Y. Zhou, S. M. Wang, *J. Alloys Compd.* 426 (2006) 375.

158. J. L. Yuan, X. Y. Zeng, J. T. Zhao, Z. J. Zhang, H. H. Chen, X. X. Yang, *J. Phys. D. Appl. Phys.* 41 (2008) 105406.

159. L. Yong, W. Xiantao, C. Hongmei, P. Gang, P. Yan, G. Li, Z. Leilei, Z. Guangyang, J. Yuexia, *J. Lumin.* 168 (2015) 124.

160. L. Yong, W. Xiantao, C. Hongmei, P. Yan, J. Yuexia, *Phys. B* 478 (2015) 95.

161. M. P. Ledent, *Chem. Phys. Lett.* 35 (1975) 375.

162. N. S. Sawala, K. A. Koparkar, N. S. Bajaj, S. K. Omanwar, *Optik – Int. J. Light Electron. Opt.*, 127 (2016) 4375.

163. X. T. Wei, S. Huang, Y. H. Chen, C. X. Guo, M. Yin, W. Xu, *J. Appl. Phys.* 107 (2010) 103.

164. J. Boje, H. M. Buschbaum, Zeitschrift fuer Anorganische und Allgemeine Chemie 619 (1993) 521.

165. J. F. Huang, A. W. Sleight, *J. Solid State Chem.* 104 (1993) 52.

166. I. Radosavljevic, J. S. O. Evans, A. W. Sleight, *J. Solid State Chem.* 137 (1998) 143.

167. J. Nause, B. Nemeth, *Semicond. Sci. Technol.* 20(4) (2005) S45.

168. D. Han, Y. Wang, S. Zhang, L. Sun, R. Han, S. Matsumoto, Y. Ino, *Sci. China Inf. Sci.* 55 (2012) 1441.

169. D. K. Hwang, S. H. Kang, J. H. Lim, E. J. Yang, J. Y. Oh, J. H. Yang, S. J. Park, *Appl. Phys. Lett.* 86 (2005) 222101.

170. A. Kuroyanagi, *Jpn. J. Appl. Phys.* 28 (1989) 219.

171. J. Dong, Y. Zhao, J. Shi, H. Wei, J. Xiao, X. Xu, J. Luo, et al., *Chem. Commun.* 50 (2014) 13381.

172. K. Mahmood, B. S. Swain, H. S. Jung, *Nanoscale* 6 (2014) 9127.

173. J. H. Noh, S. H. Im, J. H. Heo, T. N. Mandal, S. I. Seok, *Nano Lett.* 13 (2013) 1764.

174. J. H. Heo, S. H. Im, J. H. Noh, T. N. Mandal, C. S. Lim, J. A. Chang, Y. H. Lee, et al., *Nat. Photonics* 7 (2013) 486.

175. S. D. Stranks, G. E. Eperon, G. Grancini, C. Menelaou, M. J. P. Alcocer, T. Leijtens, L. M. Herz, A. Petrozza, H. J. Snaith, *Science* 342 (2013) 341.

176. J. Burschka, N. Pellet, S.-Jin Moon, R. Humphry-Baker, P. Gao, M. K. Nazeeruddin, M. Grätzel, *Nature* 499 (2013) 316.

177. M. M. Lee, J. Teuscher, T. Miyasaka, T. N. Murakami, H. J. Snaith, *Science* 338 (2012) 643.

178. J. M. Ball, M. M. Lee, A. Hey, H. J. Snaith, *Jpn. J. Appl. Phys. Sci.* 6 (2013) 1739.
179. M. Liu, M. B. Johnston, H. J. Snaith, *Nature* 501 (2013) 395.
180. O. Malinkiewicz, A. Yella, Y. H. Lee, G. M. Espallargas, M. Grätzel, M. K. Nazeeruddin, H. J. Bolink, *Nat. Photon.* 8 (2014) 128.
181. J. Qui, Y. Qiu, K. Yan, M. Zhong, H. Mu, Ch. Yan, S. Yang, *Nanoscale* 8 (2013) 3245.
182. H. Zhou, Q. Chen, G. Li, S. Luo, T.-B. Song, H.-S. Duan, Z. Hong, J. You, Y. Liu, Y. Yang, *Science* 345 (2014) 542.
183. Q. Zhang, C.S. Dandeneau, X. Zhou, G. Cao, *Adv. Mater.* 21 (2009) 4087.
184. J. Yang, B. D. Siempelkamp, D. Liu, T. L. Kelly, *ACS Nano* 9 (2) (2015) 1955.
185. R. J. Cava, B. Batlogg, R. B. van Dover, D. W. Murphy, S. Sunshine, T. Siegrist, J. P. Remeika, E. A. Rietman, S. Zahurak, G. P. Espinosa, *Phys. Rev. Lett.* 58 (1984) 1676.
186. K. I. Kobayashi, T. Kimura, H. Sawada, K. Terakura, Y. Tokura, *Nature* 395 (1998) 677.
187. S. D. Guo, B. G. Liu, *Phys. B* 408 (2013) 110.
188. E. Burzona, I. Balasz, M. Valeanu, I. G. Pop, *J. Alloys Compd.* 509 (2011) 105.
189. T. Wei, Y. Ji, X. Meng, Y. L. Zhang, *Electrochem. Commun.* 10 (2008) 1369.
190. D. Marrero-Lopez, J. Pena-Martinez, J. C. Ruiz-Moraless, *J. Solid State Chem.* 182 (2009) 1027.
191. H. Zhou, Q. Chen, G. Li, S. Luo, T. B. Song, H. S. Duan, Z. Hong, J. You, Y. Liu, Y. Yang, Interface engineering of highly efficient perovskite solar cells, *Science* 354 (2014) 542.
192. J. M. Ball, M. M. Lee, A. Hey, H. J. Snaith, *Energy Environ. Sci.* 6 (2013) 1739.
193. B. Cai, Y. Xing, Z. Yang, W. H. Zhang, J. Qiu, *Energy Environ. Sci.* 6 (2013) 1480.
194. Q. Wang, Y. Shao, Q. Dong, Z. Xiao, Y. Yuan, J. Huang, *Energy Environ. Sci.* 7 (2014) 2359.
195. C. Zuo, L. Ding, *Nanoscale* 6 (2014) 9935.
196. S. Lv, L. Han, J. Xiao, L. Zhu, J. Shi, H. Wei, Y. Xu, et al., *Chem. Commun.* 50 (2014) 6931.
197. P. Jackson, D. Hariskos, R. Wuerz, W. Wischmann, M. Powalla, *Phys. Status Solidi RRL* 8 (2014) 219.
198. P. Jackson, D. Hariskos, E. Lotter, S. Paetel, R. Wuerz, R. Menner, W. Wischmann, M. Powalla, *Prog. Photovolt.* 19 (2011) 894.
199. P. Reinhard, A. Chirila, P. Blosch, F. Pianezzi, S. Nishiwaki, S. Buecheler, A. N. Tiwari, *IEEE J. Photovolt.* 3 (2013) 572.
200. J. Britt, C. Ferekides, *Appl. Phys. Lett.* 62 (1993) 2851.
201. A. Gupta, V. Parikh, A. D. Compaan, *Sol. Energy Mater. Sol. Cells* 90 (2006) 2263.
202. A. Romeo, G. Khrypunov, F. Kurdesau, M. Arnold, D. L. Bätzner, H. Zogg, A. N. Tiwari, *Sol. Energy Mater. Sol. Cells* 90 (2006) 3407.
203. M. Gloeckler, I. Sankin, Z. Zhao, *IEEE J. Photovolt.* 3 (2013) 1389.
204. W. A. Laban, L. Etgar, *Energy Environ. Sci.* 6 (2013) 3249.
205. C. Sun, Y. Guo, H. Duan, Y. Chen, Y. Guo, H. Li, H. Liu, *Sol. Energy Mater. Sol. Cells* 143 (2015) 360.
206. J. You, Z. Hong, Y. M. Yang, Q. Chen, M. Cai, T. B. Song, C. C. Chen, et al., *ACS Nano* 8 (2014) 1674.
207. V. Gonzalez Pedro, E. J. Juarez Perez, W. S. Arsyad, E. M. Barea, F. Fabregat Santiago, I. MoraSero, J. Bisquert, *Nano Lett.* 14 (2014) 888.
208. J. Seo, S. Park, Y. Chan Kim, N. J. Jeon, J. H. Noh, S. C. Yoon, S. I. Seok, *Energy Environ. Sci.* 7 (2014) 2642.

209. H. S. Jung, N. G. Park, *Small* 11 (2015) 10.
210. W. J. Yin, T. Shi, Y. Yan, *Adv. Mater.* 26 (2014) 4653.
211. J. W. Lee, T. Y. Lee, P. J. Yoo, M. Grätzel, S. Mhaisalkar, N. G. Park, *J. Mater. Chem. A* 2 (2014) 9251.
212. J. M. P. J. Verstegen, A. L. N. Stevels, *J. Lumin.* 9 (1974) 406.
213. J. M. P. J. Verstegen, J. L. Sommerdijk, A. Bril, *J. Lumin.* 9 (1974) 420.
214. J. M. P. J. Verstegen, *J. Electrochem. Soc.* 121 (1974) 1623.
215. A. L. N. Stevels, J. M. P. J. Verstegen, *J. Lumin.* 14 (1976) 207.
216. A. L. N. Stevels, A. D. M. Schrama-de Pauw, *J. Electrochem. Soc.* 123 (1976) 691.
217. A. L. N. Stevels, *J. Lumin.* 17 (1978) 121.
218. B. Smets, J. Rutten, G. Hoeks, J. Verlijsdonk, *J. Electrochem. Soc.* 136 (1989) 2119.
219. R. Roy, D. Ravichandran, W. B. White, *J. Soc. Inf. Display* 4 (1996) 183.
220. D. Ravichandran, R. Roy, W. B. White, S. Erdei, *J. Mater. Res.* 12 (1997) 819.
221. J. M. P. J. Verstegen, J. L. Sommerdijk, J. G. Verriet, *J. Lumin.* 6 (1973) 425.
222. J. L. Sommerdijk, J. M. P. J. Verstegen, *J. Lumin.* 9 (1974) 415.
223. F. P. Glasser, L. S. D. Glasser, *J. Am. Ceram. Soc.* 46 (1963) 377.
224. S. Ito, S. Banno, K. Suzuki, M. Inagaki, *Physik. Chem.* 105 (1977) 173.
225. K. Kato, H. Saalfeld, Acta Crystallograpghy, Section B33 (1977) 1596.
226. A. J. Lindop, C. Mathews, D. W. Goodwin, *Acta Cryst.* B31 (1975) 2940.
227. S. Kimura, E. Bannai, I. Shindo, *Mater. Res. Bull.* 17 (1982) 290.
228. N. Iyi, Z. Inque, S. Takekawa, S. Kimura, *J. Solid State Chem.* 52 (1984) 66.
229. J.-G. Park, A. N. Cormack, *J. Solid State Chem.* 121 (1996) 278.
230. H. J. Song, D. H. Chen, W. J. Tang, Y. H. Peng, *Displays* 29 (2008) 41.
231. A. Nag, T. R. N. Kutty, *J. Alloys Compd.* 354 (2003) 221.
232. J. Holsa, H. Jungner, M. Lastusaari, *J. Alloys Compd.* 323 (2001) 326.
233. C. Chen, B. Wu, A. Jiang, G. You, *Sci. Sinica B* 28 (1985) 235.
234. C. Chen, Y. Wu, A. Jiang, B. Wu, G. You, R. Li, S. Lin, *J. Opt. Soc. Am. B* 6 (1989) 616.
235. Y. Wu, T. Sasaki, S. Nakai, A. Yokotani, H. Tang, C. Chen, *Appl. Phys. Lett.* 62 (1993) 2614.
236. T. Sugawara, R. Komatsu, S. Uda, *Solid State Commun.* 107 (1998) 233.
237. H. G. Giesber, J. Ballato, W. T. Pennington, J. W. Kolis, *Inf. Sci.* 149 (2003) 61.
238. H. R. Jung, B. M. Jin, J. W. Cha, J. N. Kim, *Mater. Lett.* 30 (1997) 41.
239. N. I. Leonyuk, E. V. Koporulina, V. V. Maltsev, O. V. Pilipenko, M. D. Melekhova, A. V. Mokhov, *Opt. Mater.* 26 (2004) 443.
240. C. T. Chen, B. C. Wu, A. D. Jiang, G. M. You, *Sci. Sinica B* 28 (1985) 235.
241. H. Hellwig, J. Liebertz, L. Bohaty, *Solid State Commun.* 109 (1998) 249.
242. B. Teng, J. Y. Wang, Z. P. Wang, X. B. Hu, H. D. Jiang, H. Liu, X. F. Cheng, S. M. Dong, Y. G. Liu, Z. S. Shao, *J. Cryst. Growth* 233 (2001) 282.
243. A. A. Kaminskii, P. Becker, L. Bohaty, K. Ueda, K. Takaichi, J. Hanuza, M. Maczka, H. J. Eichler, G. M. A. Gad, *Opt. Commun.* 206 (2002) 179.
244. J. Barbier, N. Penin, A. Denoyer, L. M. D. Cranswick, *Solid State Sci.* 7 (2005) 1055.
245. A. V. Egorysheva, Y. F. Kargin, *Russ. J. Inorg. Chem.* 51 (2006) 1106.
246. A. V. Egorysheva, V. M. Skorikov, V. D. Volodin, O. E. Myslitskii, Y. F. Kargin, *Russ. J. Inorg. Chem.* 51 (2006) 1956.
247. R. S. Bubnova, S. V. Krivovichev, S. K. Filatov, A. V. Egorysheva, Y. F. Kargin, *J. Solid State Chem.* 180 (2007) 596.
248. Z. Yang, X. L. Chen, J. K. Liang, Y. C. Lan, T. Xu, *J. Alloys Compd.* 319 (2001) 247.
249. S. Y. Zhang, L. Wu, X. L. Chen, M. He, Y. G. Cao, Y. T. Song, D. Q. Ni, *Mater. Res. Bull.* 38 (2003) 783.

250. Y. Zhang, X. L. Chen, J. K. Liang, T. Xu, *J. Alloys Compd.* 348 (2003) 314.
251. X. Z. Li, C. Wang, X. L. Chen, H. Li, L. S. Jia, L. Wu, Y. X. Du, Y. P. Xu, *Inorg. Chem.* 43 (2004) 8555.
252. C. T. Chen, B. Wu, A. Jiang, G. You, *Sci. China B.* 18 (1985) 235.
253. C. T. Chen, Y. Wu, A. Jiang, G. You, R. Li, S. Lin, *J. Opt. Soc. Am. B* 6 (1989) 616.
254. T. Sasaki, Y. Mori, M. Yoshimura, Y. K. Yap, T. Kamimura, *Mater. Sci. Eng. R* 30 (2000) 1.
255. P. Becker, *Adv. Mater.* 10 (1998) 979.
256. H. Zhang, G. Wang, L. Guo, A. Geng, Y. Bo, D. Cui, Z. Xu, et al., *Appl. Phys. B* 93 (2008) 323.
257. M. Yoshimura, Y. Mori, Z. G. Hu, T. Sasaki, *Opt. Mater.* 26 (2004) 421.
258. Z. W. Pei, Q. Su, J. Y. Zhang, *J. Alloys Compd.* 198 (1993) 51.
259. A. Diaz, D. A. Keszler, *Chem. Mater.* 9 (1997) 2071.
260. R. Sankar, *Solid State Sci.* 10 (2008) 1864.
261. C. J. Duan, W. F. Li, J. L. Yuan, J. T. Zhao, *J. Alloys Compd.* 458 (2008) 536.
262. Z. Y. Zhang, Y. H. Wang, J. C. Zhang, *Mater. Lett.* 62 (2008) 846.
263. P. L. Li, Z. P. Yang, Z. J. Wang, Q. L. Guo, *Mater. Lett.* 62 (2008) 1455.
264. H. Zhang, J. Chen, H. Guo, *J. Rare Earths* 29 (2011) 822.
265. K. Mostafavi, M. Ghahari, S. Baghshahi, A. M. Arabi, *J. Alloys Compd.* 555 (2013) 62.
266. K. P. Sanosh, A. Balakrishnan, L. Francis, T. N. Kim, *J. Alloys Compd.* 495 (2010) 113.
267. A. J. J. Bos, *Nucl. Instrum. Meth. B* 184 (2001) 3.
268. C. Kosanovi, N. Stubicar, N. Tomasic, V. Bermanec, M. Stubicar, *J. Alloys Compd.* 389 (2005) 306.
269. M. Kharaziha, M. H. Fathi, *Ceram. Int.* 35 (2009) 2449.
270. L. Lin, M. Yin, C. Shi, W. Zhang, *J. Alloys Compd.* 455 (2008) 327.
271. R. Naik, S. C. Prashantha, H. Nagabhushana, S. C. Sharma, B. M. Nagabhushana, H. P. Nagaswarupa, H. B. Premakumar, *Sens. Actuators B Chem.* 195 (2014) 140.
272. R. Naik, S. C. Prashantha, H. Nagabhushana, H. P. Nagaswarupa, K. S. Anantharaju, S. C. Sharma, B. M. Nagabhushana, H. B. Premkumar, K. M. Girish, *J. Alloys Compd.* 617 (2014) 69.
273. H. Yang, J. Shi, M. Gong, K. W. Cheah, *J. Lumin.* 118 (2006) 257.
274. B. N. Lakshminarasappa, S. C. Prashantha, F. Singh, *Curr. Appl. Phys.* 11 (2011) 1274.
275. S. C. Prashantha, B. N. Lakshminarasappa, F. Singh, *J. Lumin.* 132 (2012) 3093.
276. J. C. Zhang, Y. Z. Long, X. Wang, C. N. Xu, *J. Electrochem. Soc.* 4 (2014) 40665.
277. T. Wang, Y. Hu, L. Chen, *J. Mater. Sci.* 26 (2015) 5360.
278. S. A. Koparkar, N. S. Bajaj, S. K. Omanwar, *J. Rare Earths* 33 (2015) 486.
279. Y. Wang, G. Zhu, S. Xin, Q. Wang, Y. Li, Q. Wu, C. Wang, X. Wang, X. Ding, W. Geng, *J. Rare Earths* 33 (2015) 1.
280. www.mindat.org/min-3662.html Mindat.org.
281. D. L. Perry, $2Al_2O_3.SiO_2$, Al_4SiO_8 "2:1 mullite". *Handbook of Inorganic Compounds.* Taylor & Francis, ISBN 978-1-4398-1461-1 (2011).
282. J. F. Shackelford, R. H. Doremus. *Jump up to Ceramic and Glass Materials: Structure, Properties and Processing.* Springer, ISBN 978-0-387-73361-6 (2008).
283. H. Takasaki, S. Tanabe, T. Hanada, *J. Ceram. Soc.* 104 (1996) 322.
284. H. Yamamoto, T. Matsuzawa, *J. Lumin.* 73 (1997) 287.

285. M. Mueller, R. Mueller, G. Krames, M. R. Hoeppe, H. A. Stadler, F. Schnick, W. Juestel, T. Schmidt, *Phys. Status Solidi A* 202 (2005) 1727.
286. Y. Xu, D. Chen, *Ceram. Int.* 34 (2008) 2117.
287. W. Pan, G. Ning, X. Zhang, J. Wang, Y. Lin, J. Ye, *J. Lumin* 128 (2008) 1975.
288. S. Kubota, M. Shimada, *Appl. Phys. Lett.* 81 (2002) 2749.
289. S. Kubota, H. Yamane, M. Shimada, *Chem. Mater.* 14 (2002) 4105.
290. J. S. Wang, E. M. Vogel, E. Snitzer, *Opt. Mater.* 3 (1994) 187.
291. Y. Wang, J. Ohwaki, *Appl. Phys. Lett.* 63 (1993) 3268.
292. A. Lucca, M. Jacquemet, F. Druon, F. Balembois, P. Georges, P. Camy, J. L. Doualan, R. Moncorgé, *Opt. Lett.* 29 (2004) 1879.
293. R. Mueller-Mach, G. Mueller, M. R. Krames, H. A. Hoppe, F. Stadler, W. Schnick, *Phys. Status Solidi. A.* 202 (2005) 1727.
294. P. F. Smet, D. Poelman, M. P. Hehlen, *Opt. Mater. Express* 2 (2012) 452.
295. K. Inoue, N. Hirosaki, R. J. Xie, T. Takeda, *J. Phys. Chem. C.* 113 (2009) 9392.
296. R. J. Xie, H. T. Hintzen, *J. Am. Ceram. Soc.* 96 (2013) 665.
297. R. Mueller-Mach, G. Mueller, M. R. Krames, H. A. Hoppe, F. Stadler, W. Schnick, T. JuEstel, P. Schmidt, *Phys. Status Solidi A. Appl. Mater.* 202 (2005) 1727.
298. J. H. Burroghes, D. C. Bradley, A. R. Brown, R. N. Marks, K. Mackay, R. H. Friend, P. L. Burn, A. B. Holmes, *Nature* 347 (1990) 539.
299. M. C. J. M. Vissenberg, PhD Thesis, Leiden (1999).
300. M. D. McGehee, *Synth. Met.* 119-1 121 (2001) 1.
301. M. Berggren, G. Gustafsson, and O. Inganäs, M. R. Andersson, O. Wennerström, T. Hjertberg, *Appl. Phys. Lett.* 65 (1994) 1489.
302. D. Nath, P. Banerjee, *Environ. Toxicol. Pharmacol.* 36 (2013) 997.
303. V. Singh, R. P. S. Chakradhar, J. L. Rao, D. K. Kim, *J. Lumin.* 129 (2009) 130.
304. V. V. Atuchin, N. F. Beise, E. N. Galashov, E. M. Mandrik, M. S. Molokeev, A. P. Yelisseyev, A. A. Yusuf, Z. Xia, *ACS Appl. Mater. Interfaces* 7 (2015) 2623.
305. X. Dai, Z. Zhang, Y. Jin, Y. Niu, H. Cao, X. Liang, L. Chen, J. Wang, X. Peng, *Nature* 515 (2014) 96.
306. E. Jang, S. Jun, H. Jang, J. Lim, B. Kim, Y. Kim, *Adv. Mater.* 22 (2010) 3076.
307. K. Robert, S. Maria, S. Helmut, *Nat. Mater.* 13 (2014) 233.
308. Y. Ohmori, M. Uchida, K. Muro, K. Yoshino, Jpn. J. Appl. Phys. **30**, (1991) L1938.
309. D. Braun, G. Gustafsson, D. McBranch, A. J. Heeger, J. Appl. Phys. **72**, (1992) 564.
310. H. Zhu, Y. Fu, F. Meng, X. Wu, Z. Gong, Q. Ding, M. V. Gustafsson, M. T. Trinh, S. Jin, X. Y. Zhu, *Nat. Mater.* 14 (2015) 636.
311. X. Wang X. Yan, W. Li, S. Kang, *Adv. Mater.* 24 (2012) 2742.

3

Synthesis and Characterization of Phosphors

3.1 Synthesis of Phosphor Materials

The quest for new materials, methods, and technologies is currently a very active area of research in the development of new and more efficient organic and inorganic phosphors or semiconductor materials for solar cell and display technology due to their specific advantages, such as lower power consumption than incandescent light sources and better light conversion efficiency. Currently, white light–emitting diodes (WLEDs) are being incorporated into a variety of applications, stimulating the search for new materials as possible candidates for different applications. The synthetic techniques amenable to preparing phosphor materials are listed in Table 3.1.

In this chapter, we mainly focus on the solid-state diffusion, co-precipitation, sol-gel, and combustion methods, which are the most popular synthesis techniques, used for preparation of any shape and size product from the last few decades.

3.1.1 Solid-State Method

The solid-state method is most widely used for the synthesis of inorganic luminescent materials. In this method, individual metal oxides, carbonates, or nitrates are taken as starting materials to obtain the desired phase, and some flux is used to increase the rate of reaction, because this method requires a higher temperature than other conventional methods used for phosphor synthesis. The weights of the individual ingredients and doping ions are calculated in a stoichiometric ratio using concepts of chemistry. Stoichiometry is used to measure these quantitative relationships and to determine the number of products/reactants that are produced in a given reaction. Also, it describes the quantitative relationships among substances as they participate in chemical reactions; it is also called *reaction stoichiometry*. In this method, a temperature range of the order of 1200°C–1800°C is required, depending on the phosphor host, for a long time. Thus, the rate of a solid-state reaction depends on the reaction conditions, structural properties of the reactants, surface area of the solids, their reactivity, and so on.

TABLE 3.1

List of Various Methods for Preparing the Samples

Technique	Methods
Wet chemical	Co-precipitation
	Re-crystallization
Solid state	Melting
	Flux melting
	Solid-state diffusion
Novel syntheses	Spray and polymer pyrolysis
	Solid-state metathesis
	Sonochemical
	Solvothermal
	Sol-gel
	Combustion

Here, all the solid inorganic reactants are used as starting materials to obtain the desired phase. Generally, the reactants are in the form of oxides, carbonates, or nitrates. An increase in surface area enhances the rate of reaction, and fine grain–sized materials should be used. After the weights of all ingredients have been calculated, they are crushed and mixed well for better homogeneity with the help of an agate mortar and pestle. A sufficient amount of some volatile organic liquid, such as acetone or alcohol, is added to the mixture for better homogeneity. Finally, this forms a paste, which is then mixed thoroughly. During the crushing of all ingredients, the organic liquid gradually volatizes and has usually evaporated completely after 10–15 min. Also, processing by ball mills is used for better homogeneity, but this takes a long time. Finally, the ground mixture of all the ingredients is transferred into a silica or alumina crucible and kept in a furnace maintained at a constant temperature. This reaction may be carried out in different surroundings, such as a gaseous or open atmosphere. Many types of host materials, such as oxides, aluminates, silicates, tungstates, borates, vanadates, and so on, have been synthesized by solid-state methods [1–3].

3.1.2 Sol-Gel Method

The sol-gel process is another wet chemical method for producing nanomaterials. This method is used for the fabrication of metal oxide nanoparticles, especially the oxides of silicon and titanium. This method involves the conversion of monomer into a colloidal sol (solution), which acts as a precursor for an integrated network (gel) of discrete particles. The precursors used are metal alkoxides. In this process, the sol is gradually converted into gel containing both the liquid phase and the solid phase, whose morphology ranges from discrete particles to a continuous polymer network. The starting materials form a chemical solution leading to the formation of a colloidal suspension known as a *sol*. Then, the sol evolves toward the formation of an inorganic network containing a liquid phase called a *gel*. The remaining liquid phase (solvent) is removed by a drying process, and then the sol yields the gel. The particle size and shape are controlled by the sol/gel phase transition. After this, thermal treatment is provided to the gel, which is necessary for further polycondensation and to enhance the structural stability of the product. One of the advantages of this method is that it requires a lower temperature than traditional material processing methods [4–6]. The precursors used for the synthesis of colloidal nanoparticles are metal alkoxides and metal chlorides. The colloidal system, composed of solid particles dispersed in a solvent, contains particles ranging in size from 1 nm to 1 mm. The prepared sol can be further processed to obtain the substrate in a film, either by dip-coating to form a desired shape or powdered by calculations. This is an interesting, cheap, and low-temperature synthesis technique, which is used to produce a wide range of nanoparticles with controlled chemical composition. The sol-gel process is used for the fabrication of both glassy and ceramic materials, especially for aluminates, silicates, tungstates, and so on.

3.1.3 Combustion Method

Generally, for the synthesis and development of new and advanced multifunctional phosphor materials, novel synthesis techniques are required. The combustion method is an ideal synthesis technique used to obtain fine, chemically homogeneous, pure, single-phase powders in the as-synthesized condition. The combustion synthesis (CS) is a form of exothermic reaction, which is initiated at the ignition temperature and generates heat of the order of above 1000 K during the combustion, which can volatilize low–boiling point impurities and therefore results in purer and more homogeneous products than those obtained by the conventional techniques mostly used in the last few years. Nanosized particles may be achieved by this method using different fuels such as urea, tetraformal trisazine (TFTA), carbohydrazide (CH), glycine, malonodihydrazide (MDH), 3-methyl pyrazole 5-one ($3MP_5O$), dofirmyl hydrazine, and so on. During the combustion process, a large amount of gas is liberated with the occurrence of yellowish flame, and finally, a white phosphor product is obtained. The highest photoluminescence emission intensity is correlated with the highest measured flame temperature, which shows that thermal effects play an important role in achieving high-purity and high–intense phosphor powders. This is also the simplest and cheapest technique to obtain any shape and sized product within a short time. In a combustion process, metal nitrate acts as an oxidizing reactant and urea as a reducing reactant, the selection of fuel and oxidizer is an important parameter for the synthesis of different host materials. Hence, the combustion process is a complex sequence of exothermic chemical reactions between a fuel and an oxidant, accompanied by the production of heat and light in the form of yellow-ish flames. The direct combustion process under the open atmosphere is a reaction mediated by radical intermediates. Therefore, the conditions for radical production are naturally provided by thermal runaway, in which the heat generated by combustion is necessary to maintain the high temperature for radical production. The weights of the starting materials, such as metal nitrate and urea, are calculated on the basis of total oxidizing and reducing valences of the oxidizer and the fuel using the concepts of propellant chemistry [7–10].

3.1.3.1 Advantages of Combustion Method

1. The high reaction temperature can volatilize low–boiling point impurities and finally results in high-purity phosphor products.
2. The simple exothermic nature of the self-propagating high-temperature synthesis (SHS) reaction avoids the need for expensive experimental and equipment facilities.
3. The short reaction (exothermic) time results in low operating and processing costs.
4. The high thermal gradients and rapid cooling rates can be consolidated to produce a final product in one step by using the chemical energy of the reactants.
5. During complete combustion, gaseous products are released, including N_2, CO_2, and H_2O, making this an environmentally clean processing technique.
6. The larger amount of gas evolved during combustion results in a porous and homogeneous phosphor product, in which the agglomerates formed are so weak that they are easily crushed and ground into a fine powder.
7. A high temperature is formed during the chemical reaction, whose ignition temperature is lower than the crystallization temperature of the product. Hence, there is a substantial economic benefit in the production of the phosphor by SHS techniques. Due to this, it is widely used in the synthesis of various inorganic compounds.

3.1.4 Precipitation Method

This is also a simple synthesis technique for the preparation of phosphor materials. In this method, the molecular motions and the chemical reaction may proceed very swiftly in the liquid state. However, the solvent molecules may remain as contaminants, and this can adversely affect the performance of the materials. In this technique, the starting materials are dissolved in distilled water or another solvent individually and mixed well in a beaker. After simultaneous heating and stirring at low temperature, a white precipitate is formed. It is filtered, collected, dried, and finally annealed in air at high temperature for several hours to obtain the desired phase of the materials [11–13]. This technique is very useful for the synthesis of phosphate- and sulfate-based phosphor materials.

3.2 Characterization of Phosphors

Rare earth–doped inorganic luminescent materials have been characterized by different techniques, such as X-ray diffraction (XRD), scanning electron microscopy (SEM), transmission electron microscopy (TEM), high-resolution transmission electron microscopy (HRTEM), photoluminescence (PL), thermoluminescence (TL), diffuse reflectance spectroscopy (DRS), and decay time measurement.

3.2.1 X-Ray Diffraction (XRD)

XRD is a useful characterization technique to identify the phase purity and crystallinity of the material. It is also used to determine the thickness of thin film, grain size, and multilayers. It is helpful to measure structural properties such as phase composition, preferred orientation, and defect structures of crystalline or polycrystalline materials. In this technique, a monochromatic X-ray beam is incident on the sample surface, and hence, depending on the incidence angle, X-ray wavelength, and atomic spacing in the sample, the phenomenon of constructive interference is observed in the diffracted beam. Thus, when an atom of a crystal is exposed to the incident X-ray beam, electric forces are experienced due to the interaction of the charged particle with atoms of the crystal. Therefore, the atomic electrons vibrate harmonically at the frequency of the incident radiation and undergo acceleration. These accelerated charges radiate electromagnetic energy at the frequency of vibration, that is, at the incident wave frequency. In the X-ray diffraction technique, the wavelength of the incident X-ray beam is comparable to the interatomic spacing between the planes, and hence, we get an equidistant diffraction pattern of an atom. This phenomenon works on the principle of Bragg's law, $2d \sin \theta = n\lambda$, where n is an integer, λ is the wavelength, d is the crystal lattice spacing, and θ is the angle of incidence. Figure 3.1 shows the internal arrangement of an X-ray diffractometer. The observed diffraction pattern of the experimental materials is then matched with Joint Committee on Powder Diffraction Spectra (JCPDS) data to confirm the crystalline phases and purity of the materials. Figure 3.2 shows the X-ray diffraction pattern of $Sr_2Al_2SiO_7$ phosphor synthesized by the combustion method [14]. The diffraction pattern was recorded using a PAN-analytical X-ray diffractometer

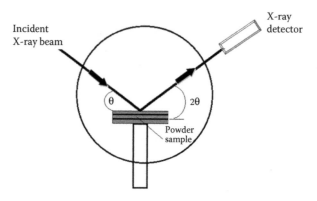

FIGURE 3.1
Internal geometry of X-ray diffractometer.

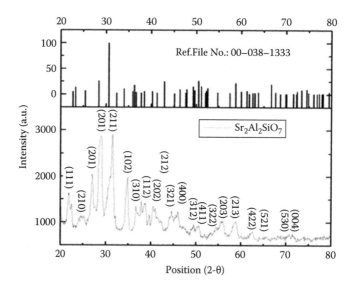

FIGURE 3.2
X-ray diffraction pattern of $Sr_2Al_2SiO_7$ phosphor. (With kind permission from Springer Science+Business Media: *Applied Physics A*, 122, 2016, 1–5, M. Kolte, V. B. Pawade, S. J. Dhoble.) [14]

with Cu–kα(1.54060 nm) radiation with step size 2h(deg.) 0.0190, scan step time (s) 31.8152, and measurement temperature (25.00°C). It is in good agreement with standard JCPDS file no.38-1333.

3.2.2 Scanning Electron Microscopy (SEM)

SEM is used to examine the surface morphologies of the sample at very high magnifications. The magnification can be greater than 3,000,000 ×. SEM analysis is a type of "non-destructive" testing of materials; that is, X-rays generated by electron interactions do not disturb the interatomic structure of the sample. Hence, it is helpful in analyzing the same sample repeatedly.

The basic components of an SEM instrument are

- Electron source (gun)
- Electron lenses
- Sample stage
- Detectors for all signals of interest
- Display/data output devices
- Infrastructure requirements:
  - Power supply
  - Vacuum system
  - Cooling system
  - Vibration-free floor
  - Room free of ambient magnetic and electric fields

The main features of the JEOL 6380A SEM model are that it consists of an electron column containing an electron source (i.e., gun), the magnetic focusing lenses, the sample vacuum chamber and stage region (at the bottom of the column), and the electronics console containing the control panel, the electronic power supplies, and the scanning modules. It is attached to an electron diffraction scattering (EDS) to identify the quantitative composition of the element in the sample.

3.2.2.1 Fundamental Principles of SEM

In SEM, the accelerated electrons carry a significant amount of kinetic energy, and this is dissipated as a variety of signals produced during the electron–materials interactions when the incident electrons are deaccelerated in the solid sample. In SEM analysis, a beam of electrons is focused on the specimen. The electron beam has energies that range from a few to 50 keV, and it is focused by two condenser lenses into a beam with a very fine spot size. This beam is then passed through the objective lens and deflected by the pairs of coils, which are arranged linearly or in a raster fashion over a rectangular area of the sample surface, as shown in Figure 3.3. When the primary electron beam strikes the surface of the sample specimen, first, it is scattered by atoms in the sample. Through the scattering process, the primary beam spreads effectively and fills a teardrop-shaped volume, known as the *interaction volume*, which extends about 1–5 μm into the surface. Secondary electrons emitted from interactions in this region are then detected, converted, and amplified to produce an image. The schematic of SEM is shown in Figure 3.3.

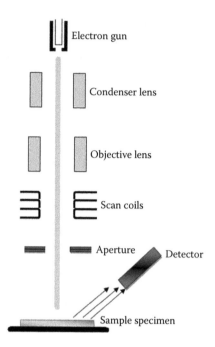

FIGURE 3.3
Schematic diagram of a scanning electron microscope.

SEM can produce more than one type of signal, including secondary electrons, backscattered electrons, characteristic X-rays, cathodoluminescence, specimen current, and transmitted electrons. The secondary and backscattered electrons are conventionally separated according to their energies, which are mostly used for imaging samples. The secondary and backscattered electrons are more valuable to show morphology and topography and also illustrate contrasts in the composition in multiphase samples. An SEM image of $Sr_2Al_2SiO_7$ phosphor powder synthesized by the combustion method is shown in Figure 3.4. From this SEM micrograph, it is seen that the particles have an irregular shape, and the multi-grains agglomerate together forming clusters with an average size in the order of about 3–5 μm. This SEM micrograph of the prepared sample was produced using a JEOL - JSM 6380A SEM instrument.

3.2.3 Transmission Electron Microscopy (TEM)

The first transmission electron microscope was built by Knoll and Ruska in 1932. TEM is a microscopy technique in which a beam of electrons is transmitted through an ultra-thin specimen, interacting and passing through it. The image is formed through interactions of the electrons transmitted through the specimen. It is then magnified and detected by a sensor, such as a charge-coupled device (CCD). Finally, it is displayed on a fluorescent screen.

FIGURE 3.4
SEM micrographs of $Sr_2Al_2SiO_7$ phosphor. (With kind permission from Springer Science+ Business Media: *Applied Physics A*, 122, 2016, 1–5, M. Kolte, V. B. Pawade, S. J. Dhoble.) [14]

3.2.3.1 Construction and Operation

Figure 3.5 shows the TEM with its different working components. It is composed of

- A vacuum system
- A series of electromagnetic lenses
- Electrostatic plates

In a transmission electron microscope, the sample is illuminated by a defocused electron beam, which is transmitted through the sample specimen and then used for imaging. The condenser lenses produce the illuminating electron beam. A lens system arranged below the sample—objective lens, intermediate lens, and projector lens—is used to form the image. This arrangement can be used in two different modes: the imaging or the diffraction mode. They are selected by the excitation of the intermediate lens. It is used to magnify the first intermediate image (imaging mode) or the diffraction pattern (diffraction mode) formed by the objective lens. The first intermediate image or the diffraction pattern is then further magnified by the projector lens, and finally, it is detected by the CCD camera, and the final image is displayed on the screen. The contrast in the final TEM image occurs due to the scattering of the incident electrons by the sample specimen. Both the amplitude and the phase of the electron waves are changed during their transit through the specimen; thus, both can contribute a TEM image. Hence, a fundamental distinction is observed in TEM between amplitude contrast and phase contrast. Both the contrasts are normally involved in the image formation process. Figure 3.6 shows a TEM image of $KSr(PO_4)$ phosphor synthesized by wet chemical methods; it is seen that the material has a core-shell structure under 20 nm resolution. This image was produced using a Hitachi (H-7500) 120 kV transmission electron microscope equipped with a CCD camera with resolution of 0.36

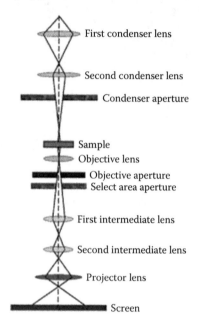

FIGURE 3.5
TEM with its components and schematic representation of TEM.

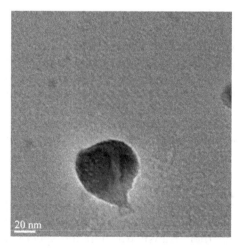

FIGURE 3.6
TEM image of $BaMg_2Al_6Si_9O_{32}$ phosphor (synthesized by combustion method).

nm (point to point) and 40–120 kV operating voltage, which can magnify a sample in high-resolution mode.

3.2.4 High-Resolution Transmission Electron Microscopy

HRTEM is a technique for high-magnification studies of nanostructured materials. High resolution makes it perfect for imaging materials on the atomic scale [15, 16]. It is a powerful technique to study the properties of materials on the atomic scale of a crystalline material, such as semiconductors, metals, nanoparticles, graphene, and carbon nanotubes. HRTEM is also referred to as *high-resolution scanning TEM* and *phase contrast TEM*. A resolution of around 0.5 Å (0.050 nm) can be achieved in phase contrast TEM [17]. In this instrument, we can characterize small-scale nanoparticles, individual atoms of a crystal, and the defects present inside the materials. This technique is also useful to know the exact structure of atoms, and hence, it is also called *electron crystallography*. The contrast of an HRTEM image arises from the phenomenon of interference in the image plane of the electron wave with itself, and the amplitude in the image plane is recorded. This gives structural information on the sample as well as d-spacing between the planes. To detect the wave, the aberrations of the microscope have to be tuned in such a way that it will convert the amplitudes in the image plane. Thus, the interaction of the electron wave with the crystallographic structure of the sample is complex, but a qualitative idea of the interaction can be readily obtained. The imaging electron from the wave interacts independently with the sample surface. While incident on the sample surface, it penetrates through it and is attracted by the positive atomic potentials of the atom cores in channels along the atom columns of the crystallographic lattice [18]. The interaction between the electron wave in different atom columns leads to Bragg's law of diffraction ($2d\sin\theta = n\lambda$). Thus, based on the dynamical theory of electron scattering, image formation by the electron microscope is sufficiently well known to allow accurate simulation of electron microscope images [18,19]. Therefore, HRTEM is an important tool used for nanomaterials to characterize the lattice structure and other useful information related to shape, size, defects, and so on. Figure 3.7a to f shows the HRTEM image of a $KSr(PO_4)$ phosphor material synthesized by wet chemical methods [20]. It is first characterized by TEM to determine the exact structure of $KSr(PO_4)$ phosphor crystallites, as shown in Figure 3.7a; this shows a small core-shell structure with different crystallite sizes. For further investigation, it is characterized by HRTEM as indicated in Figure 3.8b through f. Figure 3.7b and c show the HRTEM image observed under 20 nm resolution, which clearly indicates the spherical shape of phosphor nanoparticles. With 1, 2, and 5 nm resolution, as shown in Figure 3.7d, it is seen that particles have a spherical shape with the separation of equidistant lattice planes on the surface as depicted in Figure 3.7e. Here, we have estimated

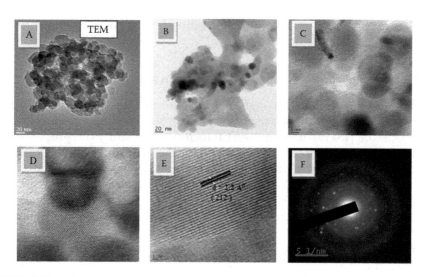

FIGURE 3.7

TEM and HRTEM images of KSr(PO₄) phosphor. (With kind permission from Springer Science+Business Media: *Journal of Materials Science: Materials in Electronics*, 28, 2017, 16306, V. B. Pawade, A. Zanwar, R. P. Birmod, S. J. Dhoble, L. F. Koao.) [20]

the spacing between the crystallographic planes, which was found to be 2.2 Å, showing good agreement with the indexed plane of (212) in the JCPDS pattern. Figure 3.7f shows the electron diffraction images of the lattice point. Hence, this is a more qualitative technique to determine the structure of the material.

3.2.5 Fluorescence Spectrophotometry

Fluorescence is defined as the emission of light by a substance that has absorbed light or some other electromagnetic radiation incident on it. It is a form of luminescence. In general, the emitted light has a longer wavelength and is lower in energy than the absorbed light. Today, the application of fluorophotometry has been expanded to various fields such as industrial materials (organic electroluminescence materials, liquid crystals), environment related (water quality analysis), pharmaceutical manufacturing (synthesis and development of a fluorescence reagent), and biotechnology (intracellular calcium concentration measurement). Among the mentioned advantages of fluorescence spectrophotometry discussed in Section 2.1, this is also an important measurement techniques to study photoluminescence behavior of organic and inorganic materials. These are recorded using, for example, a Hitachi F-4000 fluorescence spectrophotometer. This spectrophotometer basically consists of two monochromators (one on the excitation and the other on the emission side), a light source, two detectors (one for measurements and the other for monitoring), a sample holder, a data processor, and a graphics plotter.

Detailed specifications of the Hitachi F-4000 fluorescence spectrophotometer [21] are as follows:

- *Monochromators*: Large stigmatic concave gratings having 900 lines/mm are used on both excitation and emission sides with eagle mounting (F: 3). Blaze wavelengths are 300 nm on excitation side and 400 nm on emission side.
- *Measuring wavelength range*: 220–730 nm and zero order light on both excitation and emission sides.
- *Light source*: 150 W xenon lamp with ozone self-dissociation function.
- *Detector*: Photomultiplier R372 F for measurements and photoelectric tube R 518 for monitoring.
- *Resolution*: 1.5 nm (with minimum band pass).
- *Wavelength accuracy*: ±2 nm.
- *Wavelength scan rate*: 2, 5, 15, 30, 120, 240, 600 nm/min (eight-step selection).
- *Excitation side*: 1.5, 3, 5, 10, 20 nm (five steps).
- *Emission side*: 1.5, 3, 5, 10, 20, 40 nm (six steps).

3.2.5.1 Mechanism of the Spectrofluorophotometer

A block diagram of a spectrofluorophotometer is shown in Figure 3.8. Here, the flash lamp is used as an excitation source for providing excitation light to irradiate a specimen, and usually, a xenon flash lamp is used for this purpose. We know that white light composed of various wavelengths is emitted from the light source and enters an excitation side spectroscope (2). While

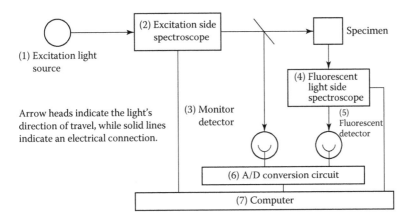

FIGURE 3.8
Block diagram of spectrofluorophotometer (Hitachi High-technologies Corporation.2001.2017). (From www.hitachihightech.com/global/products/science/tech/ana/uv/fl_basic/index.html.)

measuring an excitation spectrum, this excitation side spectroscope (2) is moved successively to change the wavelength used to irradiate the specimen (called *wavelength scanning*). In contrast, when measuring a fluorescence spectrum, a specific wavelength is selected and fixed in the excitation side spectroscope. Now, the light emitted from the excitation side spectroscope travels toward the specimen to excite it by the incidence of particular wavelengths of light radiation. Along the way, a half-mirror is used to split the light beam (as indicated by the diagonal line in the figure) so that a portion also travels to the monitor detector (3). The monitor detector monitors the intensity of the excitation light irradiating the specimen. Generally, in such apparatus, a photomultiplier tube is used. When the excitation light reaches the specimen, the specimen is excited to emit fluorescent light. For measuring the excitation spectrum, a specific wavelength is selected and fixed in the fluorescent light side spectroscope (4). When measuring a fluorescence spectrum, the fluorescent light side spectroscope is moved to measure the wavelengths of the emitted fluorescent light. The fluorescent light leaving the fluorescent light side spectroscope enters the fluorescence detector (5). Generally, a photomultiplier tube is used. A fluorescent light detector converts the fluorescent light into an analog electric signal, which is converted into a digital signal using an A/D conversion circuit (6). A computer (7) controls wavelength scanning and digital signal processing. Here, both the excitation and emission spectra are recorded within the wavelength range of 220–600 nm. Samples whose emission wavelengths are observed within the given excitation range (220–400 nm) are scanned in the correct spectrum mode, and samples whose wavelengths are beyond 600 nm are scanned in ordinary mode. Emission spectra

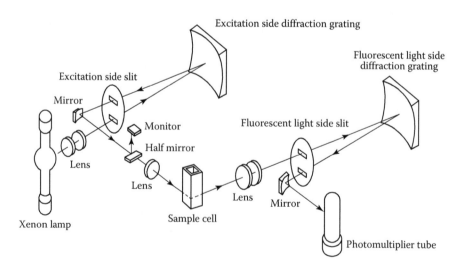

FIGURE 3.9

Optical system in spectrofluorophotometer. (From www.hitachihightech.com/global/products/science/tech/ana/uv/fl_basic/index.html.)

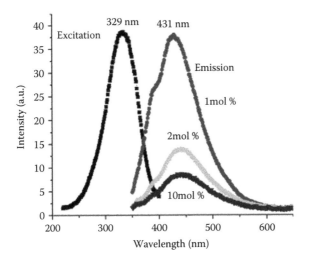

FIGURE 3.10
PLE and emission spectra of ZAM: Ce^{3+} phosphor. With kind permission from Elsevier: *Journal of Luminescence*, 135, 2013, 318, V. B. Pawade, S. J. Dhoble.) [22]

are recorded with excitation and emission band pass under 5 nm and 1.5 nm, respectively, while excitation spectra are recorded with excitation band pass at 1.5 nm and emission band at 5 nm. Figure 3.9 shows the optical system diagram of a general spectrofluorophotometer, including the different lenses used in optical pathway, the position of the sample holder, the xenon lamp, the photomultiplier tube, and so on. The recorded photoluminescence excitation and emission spectra of ZAM: Ce^{3+} inorganic phosphor materials are shown in Figure 3.10, in which the emission band is observed at 431 nm by keeping the excitation wavelength constant at 329 nm, located in the broad spectral range [22]. The *photoluminescence excitation* and emission measurements are carried out at a constant slit width (1.5 nm).

3.2.6 Decay Time Measurement

A fluorescence spectrometer is a modular, computer-controlled spectrometer for measuring the photoluminescence lifetimes of phosphor materials, spanning a vast time range of more than 12 orders of magnitude, from picoseconds to seconds. Two different single photon counting techniques are required to cover the large time range: time-correlated single photon counting (TCSPC) and multi-channel scaling (MCS).

3.2.6.1 Operating Principle

A schematic diagram of a fluorescence spectrometer is shown in Figure 3.11. In TCSPC, the sample is repetitively excited by a pulsed light source with a

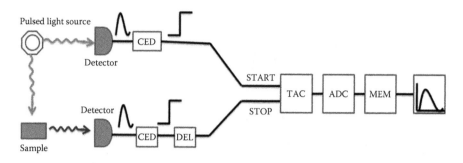

FIGURE 3.11
Schematic diagram of photoluminescence spectrometer (Edinburgh FLS980 spectrometer). (From www.edinst.com/in/products/fls-980-fluorescence-spectrometer/.) [23]

high repetition rate. During the decay time measurement, a probability histogram builds, which shows the time between an excitation pulse (START) and the observation of the first fluorescence photon (STOP).

In this instrument, the START signal triggers a linear voltage ramp of the time to amplitude converter (TAC). This ramp is stopped when the first fluorescence photon is detected. The TAC produces a voltage output, which is proportional to the time duration between the START and STOP signals. This voltage is read by an analog to digital converter (ADC), and the value is stored in the memory (MEM). Summing over many START-STOP cycles, the evolution of the probability histogram can be displayed in real time. It represents the growth and decay of the fluorescence [23]. We know that when radiation is emitted from a molecule after excitation, it exhibits a multiparameter nature. Therefore, the decay time measurement is an important parameter to gain information about the steady-state measurement of the fluorescence lifetime of a substance. The fluorescence lifetime gives an absolute measure and allows a dynamic picture of the obtained fluorescence spectra. One of the most important advantages of this technique including the use of fluorescence lifetime is that it gives an absolute measurement, unlike the steady-state intensity [24]. The fluorescence lifetime is an intrinsic molecular property independent of the concentration, within certain constraints. This means that changes in concentration, whether caused by photobleaching or diluting/concentrating the sample, would not affect the lifetime value of the materials. This technique is used to investigate the lifetime value of rare earth ions when doped into a host lattice, and it is also important to know the phosphorescence properties of phosphor materials as well as the effect of acceptor rare earth ions on the decay time of donor ions. Thus, the change in intensity of the recorded emission spectra can be observed. Further, we can study TCSPC, in which the lifetime is not influenced by fluctuations in excitation source intensity. Therefore, fluorescence is an ideal nanoscale probe, because the fluorescence decay takes place on the nanosecond timescale and can be influenced by molecular processes occurring in the nanometer range. This technique is very useful for

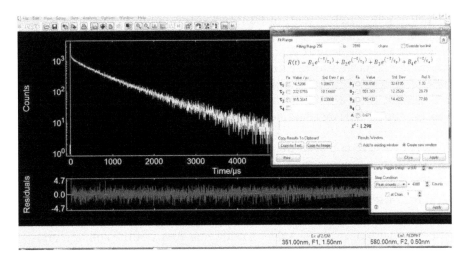

FIGURE 3.12
Fluorescence decay curve of $KNa_3Al_4Si_4O_{16}$:Ce^{3+}, Dy^{3+} phosphor. (With kind permission from Elsevier: *Journal of Physics and Chemistry of Solids* 116, 2018: 338. M.M. Kolte, V.B. Pawade, A.B. Bhattacharya, and S.J. Dhoble.) [25]

several materials that have the ability to emit light. Fluorescence lifetimes are measured at wavelength increments for the same data collection time. The resultant intensity–wavelength–time surface can then simply be "time-sliced" in the intensity–time axis to provide spectra at different times after excitation, called *time-resolved emission spectra*. Figure 3.12 shows the nature of the fluorescence decay curve of $KNa_3Al_4Si_4O_{16}$:Ce^{3+}, Dy^{3+} phosphor synthesized by combustion methods, which indicates that the prepared phosphor exhibits a fluorescence decay time of 4 µs. Also, with the help of the ratio of number of photons absorbed to number of photons emitted per second, we can calculate the quantum efficiency of the phosphor materials.

3.2.7 Diffuse Reflectance Method

Infrared spectroscopy can be used to analyze solid, liquid, or gas samples. Here, we discuss diffuse reflectance infrared Fourier transform spectroscopy (DRIFTS). This is used for the measurement of fine particles and powders as well as rough surfaces. It is a simple and easy technique for reflectance/transmittance measurement, in which no sample preparation is required for analysis.

3.2.7.1 Operation

When the infrared (IR) beam enters the sample specimen, it can be reflected off or transmitted through the surface of a particle. The IR energy reflecting off the sample surface is typically lost. The IR beam that passes through a particle is either reflected off or transmitted through the next particle. This transmission–reflectance process can occur many times in the sample, which enhances the path length. Finally, the scattered IR energy is collected by a spherical mirror and focused on the detector. The detected IR light is partially absorbed by particles and gives information about the sample.

Generally, there are three ways to prepare samples for diffuse reflectance spectroscopy (DRS) measurement.

- Fill a micro-cup with the powder. It is filled in the same way every time to maintain proper focus.
- Scratch the sample surface with a piece of abrasive paper and then measure the particles adhering to the paper.
- Place a few drops of solution on the sample substrate. Dissolve the powders in a volatile solvent, put a few drops of the solution on a substrate, evaporate it, and then analyze the remaining particles on the substrate.

Thus, DRS is a useful technique to study the optical band gap of inorganic materials. These measurements have been performed in the 190–800 nm wavelength range using a Cary-60 UV-VIS-NIR spectrophotometer with a Harrick Video-Barrelino diffuse reflectance probe. The beam spot size on the sample is around 1.5 mm in diameter, and an integrating sphere detector is used for diffuse signal detection. Diffuse reflectance is based on the focused projection of the spectrometer beam into the sample, where the beam is reflected, scattered, or transmitted through the sample. Figure 3.13 shows that incident light reflected symmetrically with respect to the normal line is called *specular reflection*, while incident light scattered in different directions is called *diffuse reflection*. Figure 3.14 shows a schematic view of the optical system in the DRS-1800, in which M_3 helps to shine IR light incident on the sample specimen, and M_4, placed exactly opposite M_3, helps to collect the diffuse reflected light, which passes through M_5 and M_6 and finally, toward the detector.

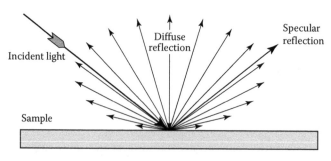

FIGURE 3.13
General reflection mechanisms in DRS.

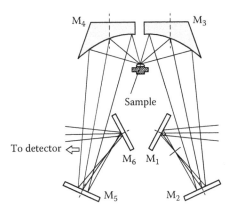

FIGURE 3.14
Schematic view of DRS-8000 diffuse reflectance optical system.

3.2.7.2 DRS Study of Phosphors

The measurement of diffuse reflection, as well as specular reflection using an integrating sphere, is carried out by keeping the sample in front of the incident light window and concentrating the light reflected from the sample onto the detector using a sphere coated inside with barium sulfate. The achieved results are the reflectance (relative reflectance) with respect to the reflectance of the reference standard. Figure 3.15 shows the reflectance (%) versus wavelength trace of $KSr(PO_4):Ce^{3+}$ inorganic nanophosphor prepared by the wet chemical route. It is seen that the drastic drop in reflection observed at 220, 262, and 336 nm corresponds to the UV range, which clearly indicates the optical band gap of the $KSr(PO_4):Ce^{3+}$ host lattice. Here, a strong absorption band of $KSr(PO_4):Ce^{3+}$ is observed at 336 nm, and one weak absorption is located at 760 nm [20]. Both the bands are assigned to the 4f–5d allowed transition of impurity ions. Above 336–750 nm, there are no absorption bands observed in the spectrum. Table 3.2 shows the values of photon energy (electronvolts) and K/F ratios, which have been calculated from the observed DRS spectra of $KSr(PO_4):Ce^{3+}$ phosphor.

3.2.7.3 Photon Energy

DRS is an excellent sampling tool for studying powdered or crystalline materials in the mid-IR and near-IR spectral region. We have calculated the optical band gap of the Ce^{3+}-doped $KSr(PO_4)$ phosphor using Table 3.2. Ce^{3+} ion shows broadband emission due to the 5d–4f allowed transition under the UV excitation wavelength. Thus, by using the numerical values of the Kubelka–Munk coefficient (K/S), we can calculate the band gap from the measured diffuse reflectance spectra.

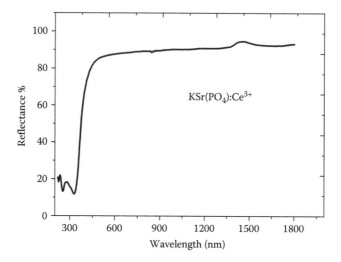

FIGURE 3.15

DRS spectra of KSr(PO4):Ce^{3+} inorganic phosphor materials. (With kind permission from Springer Science+Business Media: *Journal of Materials Science: Materials in Electronics*, 28, 2017, 16306, V. B. Pawade, A. Zanwar, R. P. Birmod, S. J. Dhoble, L. F. Koao.) [20]

TABLE 3.2

Calculation of Photon Energy and Kubelka–Munk Function

Wavelength (x) (nm)	Intensity (y) (a.u)	1240/X	F = Y/100	K = (Y/100)^2	K/F
220	20.823	5.636363636	0.20823	0.043359733	0.20823
222	19.484	5.585585586	0.19484	0.037962626	0.19484
224	18.395	5.535714286	0.18395	0.033837603	0.18395
226	18.435	5.486725664	0.18435	0.033984923	0.18435
228	19.235	5.438596491	0.19235	0.036998523	0.19235
230	20.39	5.391304348	0.2039	0.04157521	0.2039
232	21.424	5.344827586	0.21424	0.045898778	0.21424
234	21.924	5.299145299	0.21924	0.048066178	0.21924
236	21.791	5.254237288	0.21791	0.047484768	0.21791
238	21.122	5.210084034	0.21122	0.044613888	0.21122
240	20.022	5.166666667	0.20022	0.040088048	0.20022
242	18.658	5.123966942	0.18658	0.034812096	0.18658
244	17.249	5.081967213	0.17249	0.0297528	0.17249
246	15.949	5.040650407	0.15949	0.02543706	0.15949
248	14.828	5	0.14828	0.021986958	0.14828
250	13.971	4.96	0.13971	0.019518884	0.13971
252	13.459	4.920634921	0.13459	0.018114468	0.13459

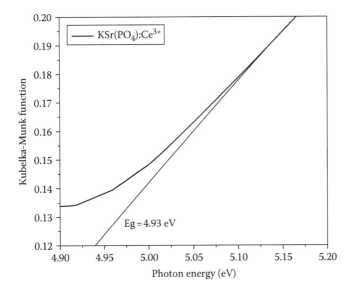

FIGURE 3.16
Plot of photon energy versus Kubelka–Munk function. (With kind permission from Springer Science+Business Media: *Journal of Materials Science: Materials in Electronics*, 28, 2017, 16306, V. B. Pawade, A. Zanwar, R. P. Birmod, S. J. Dhoble, L. F. Koao.) [20]

$$\frac{K}{S} = \frac{(1-R)^2}{2R} \qquad (3.1)$$

where:
  K = absorption coefficient
  S = scattering coefficient
  R = reflectivity

The plot of photon energy versus Kubelka–Munk function is shown in Figure 3.16. From the plot, the value of photon energy for KSr(PO$_4$):Ce^{3+} phosphor is found to be 4.93 eV. Hence, it is confirmed that KSr(PO$_4$) is a promising phosphate host due to its large band gap and may be applicable in many optical and semiconductor devices.

References

1. M. Li, L. L. Wang, W. Ran, C. Ren, Z. Song, J. Shi, *J. Lumin.* 184 (2017) 143.
2. Y. Wang, A. Chen, C. Tu, *Optical Materials* 47 (2015) 561.
3. Z. Zhu, G. Fu, Y. Yang, Z. Yang, P. Li, *J. Lumin.* 184 (2017) 96.
4. N. Elham, S. N. Masoud, B. Mehdi, G. Tahereh, *J. Clust. Sci.* 24 (2013) 1171.

5. M. R. Allahgholi Ghasri, R. Fazaeli, B. Gorji, N. Niksirat, *J. Appl. Chem. Res.* 6 (2012) 22.

6. A. Y. Abdel-Latief, M. A. Abdel-Rahim, M. N. Abdel Salam, A. Gaber, *Int. J. Electrochem. Sci.* 9 (2014) 81.

7. V. B. Pawade, S. J. Dhoble, *J. Lumin.* 145 (2014) 626.

8. V. B. Pawade, N. S. Dhoble, S. J. Dhoble, *J. Rare Earths* 32 (2014) 593.

9. V. B. Pawade, H. C. Swart, S. J. Dhoble, *Renew. Sust. Energ. Rev.* 52 (2015) 596.

10. V. B. Pawade, N. S. Dhoble, S. J. Dhoble, *Mater. Res. Express* 2 (2015) 095501.

11. K. N. Kumar, B. C. Babu, S. Buddhudu, *J. Lumin.* 161 (2015) 456.

12. H. Zhu, Y. Liu, D. Zhao, M. Zhang, J. Yang, D. Yan, C. Liu, et al., *Optic Mater.* 59 (2016) 55.

13. Y. Tian, F. Lu, M. Xing, J. Ran, Y. Fu, Y. Peng, X. Luo, *Optic. Mater.* 64 (2017) 58.

14. M. Kolte, V. B. Pawade, S. J. Dhoble, *Appl. Phys. A* 122 (2016) 1.

15. C. H. John, *Experimental High-resolution Electron Microscopy.* New York, NY: Oxford University Press. [ISBN 0-19-505405-9] (1988).

16. C. H. John, Imaging dislocation cores—the way forward. *Philos. Mag.* 86 (2006) 4781.

17. C. Kisielowski, B. Freitag, M. Bischoff, H. van Lin, S. Lazar, G. Knippels, P. Tiemeijer, et al., *Microsc. Microanal.* 14 (2008) 469.

18. P. Geuens, D van Dyck, *Ultramicroscopy* 3 (2002) 179.

19. M. A. O'Keefe, P. R. Buseck, S. Iijima, *Nature* 274 (1978) 322.

20. V. B. Pawade, A. Zanwar, R. P. Birmod, S. J. Dhoble, L. F. Koao, *J. Mater. Sci. Mater. Electron.* 28 (2017) 16306.

21. www.hitachihightech.com/global/products/science/tech/ana/uv/fl_basic/index.html

22. V. B. Pawade, S. J. Dhoble, *J. Lumin.* 135 (2013) 318.

23. FLS980-Spectrometer Edinburgh Instruments. (www.edinst.com/in/products/fls-980-fluorescence-spectrometer/).

24. Time-Resolved Fluorescence Technical Note TRFT-1, HORIBA Scientific.

25. M. M. Kolte, V. B. Pawade, A. B. Bhattacharya, S. J. Dhoble, *J. Phys. Chem. Solids* 116 (2018) 338.

26. www.shimadzu.com/an/ftir/support/ftirtalk/talk1/intro.html

Part III

Roles of Phosphors

4

Phosphors for Light-Emitting Diodes

4.1 Origins of Light-Emitting Diodes

In recent years, demand for maintainable and environment-friendly energy sources has rapidly increased. Due to this, technologies based on renewable sources, such as fuel cells and solar cells, have been extensively studied [1–4]. On the other hand, to reduce energy consumption demands, various technologies have been adopted to get similar performance. Technologies based on fossil fuel consume a large amount of energy, half of which is wasted due to inefficient conversion of energy from one form to another [5]. Lighting devices are one such technology, in which most of the energy is wasted during the production of light. Basic lighting devices, such as incandescent bulbs and fluorescent lamps, have not evolved in terms of their energy consumption problems since these technologies were invented. In comparison to other mechanical device industries, much less attention has been given to the improvement of lighting technology. Lighting alone consumes more than 20% of total energy demand. Hence, a new lighting technology, known as a *light-emitting diode* (LED), evolved to reduce the energy demand [6]. In the past several decades, digital technology has evolved considerably, which has changed many aspects of our day-to-day life. The recent advances in design and manufacturing of full-spectrum LEDs have overcome the use of energy-consuming tungsten bulbs and fluorescent lamps. Extensive continuing research on LEDs promises to produce a pure white light emission with longer LED lifetimes [7]. In the past 10–15 years, LEDs have become popular as an efficient source for white light production. The development of this technology began in the early 1990s [8], when a breakthrough enabled the fabrication of the first blue LED. Later, red-emitting LEDs opened up the possibility of creating the entire visible spectrum through multiple combinations of blue-, green-, and red-emitting LEDs. The fascinating thing about LEDs is that the light can be produced by tiny semiconductor devices, which further helped in miniaturizing lighting technology. The extensive work on LEDs by Isamu Akasaki, Hiroshi Amano, and Shuji Nakamura has lifted this technology to potential heights, for which they were awarded the Nobel Prize in the year 2014 [9].

Nowadays, LEDs are widely used in day-to-day life for general lighting and have nearly replaced incandescent bulbs and fluorescent lamps. These modern LEDs offer several benefits over conventional incandescent light. LEDs are much more efficient than incandescent bulbs in terms of converting electricity into visible light; they are compact and tiny; and most importantly, they have a lifespan of 100,000 hours, which is almost 100 times longer than the lifespan of incandescent bulbs [10, 11]. Furthermore, LEDs offer several other superior characteristics: they are monochromatic emitters of light, can produce bright light emission, and can produce any color. Due to these characteristics, LEDs are used in a number of applications in modern technology. LEDs are now being used in automotive taillights, turn signals, and side marker lights. The long lifespan of LEDs allows manufacturers to integrate the brake light into the vehicle design without the necessity of providing for frequent replacement, as is required when incandescent bulbs are used. Single color–emitting LEDs are also being used in runway lights at airports and as warning lights on radio and television transmission towers. Incandescent lights at traffic signals consume 50–150 Watts; compared with this, LEDs in traffic lights consume 10–25 Watts for a light of similar brightness.

4.1.1 Principle and Operation

A schematic of the operating principle of an LED is given in Figure 4.1. LEDs work on the principle of a p–n junction diode. A p–n junction converts the incident light energy into an electrical current, where the intensity of absorbed light is proportional to the electric current. LEDs also use the same principle, but in reverse; a p–n junction emits light when an electrical current passes through it. This phenomenon is also called *electroluminescence*. The charge carriers recombine in a forward-biased p–n junction as the

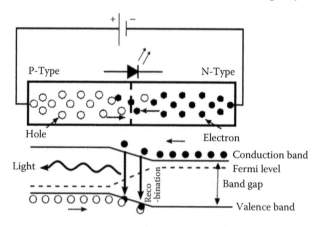

FIGURE 4.1
Schematic of carriers' recombination process in LED.

electrons cross from the n-region and recombine with the holes existing in the p-region, as shown in Figure 4.1. In the presence of an electric field, the electrons make the transition to the conduction band, leaving behind holes in the valence band. This creates an energy difference between the electrons and the holes; thus, the energy level of the holes is less than the energy level of the electrons. When the electrons from the conduction band recombine with the holes in the valence band, the resulting energy, due to the recombination of electrons and holes, is dissipated in the form of heat and light, as shown in Figure 4.1. Depending on the amount of light or heat emitted from the semiconductor, the LED can be an efficient or an inefficient emitter of light. For silicon and germanium diodes, the electrons dissipate energy in the form of heat, but in gallium arsenide phosphide (GaAsP) [12] and gallium phosphide (GaP) [13–15] semiconductors, the electrons give out energy by emitting photons. For a semiconductor to work as an LED, the semiconductor has to be translucent, and the p–n junction is used as a source of light emission. When the LED is used in reverse biased mode, there is no light emission from the junction, and if the applied potential is further increased, the device can be damaged.

4.1.2 Light Emitting Diode Colors

We know that an LED is a semiconductor device that emits light when a small voltage is applied to its terminal, it always works when it is forward biased. Light is produced when charge carriers (i.e., electrons and holes) combine together within the semiconductor material. Thus, light is coming from inside the junction on the basis of the electroluminescence mechanism. Hence, such types of devices are called *solid-state devices*, because light is generated within the solid semiconductor material. Technologies based on this type of solid-state device have gained more importance in recent years due to their long operation lifetime and better energy saving capabilities than other sources, such as incandescent and tungsten halogen lamps or gas discharge, fluorescent lamps, and so on. The emission colors of LEDs depend on the semiconductor material used, which acts as a source to excite the impurity ions that will emit the photon. Hence, in LEDs, the semiconductor material is composed of electrons and holes that are separated by energy bands. The band gap of the semiconductor materials determines the energy of the photons that are emitted by the LED, and the corresponding energy of the photons represents the emission wavelength of the emitted light; that is, the color of the LED. At present, there are many types of semiconductor materials used for the fabrication of LEDs, and every material has a specific emission wavelength; we have summarized emission wavelengths and their corresponding semiconductor materials in Table 4.1. Therefore, different semiconductor materials with different band gaps produce different colors of light. We can tune the color emission properties of LEDs by varying the composition of active or impurity ions. Most of the compound

TABLE 4.1

Emission Wavelength of Different Types of Semiconductor Materials

Color	Wavelength (nm)	Types of Semiconductor Materials
Infrared	>760	GaAs
		AlGaAs
Red	610–760	AlGaAs
		GaAsP
		AlGaInP
		GaP
Orange	590–610	GaAsP
		AlGaInP
		GaP
Yellow	570–590	GaAsP
		AlGaInP
		GaP
Green	500–570	GaInP
		AlGaP
		AlGaP
		InGaN
Blue	450–500	ZnSe
		InGaN
		SiC
		Si
Violet	400–450	InGaN
Purple	Multiple types	Dual blue/red LEDs
		Blue with red phosphor
		White with purple plastic
UV	<400	Diamond
		Boron nitride
		Aluminum nitride
		AlGaN
		AlGaInN
Pink	Multiple types	Blue + phosphor
		Yellow + red, orange, or pink phosphor
		White + pink pigment
White	Broad spectrum	Blue/UV diode + yellow phosphor

semiconductor materials that are used in LEDs belong to groups III and V of the periodic table (called *III–V materials*). Examples of III–V materials commonly used to fabricate LED devices are GaAs and GaP. In the mid-1990s, LEDs had limited colors, and at that time, blue- and white-emitting LEDs did not exist. Later, the discovery of new semiconductor materials, such as gallium nitride–based materials, helped to develop different colors of LEDs and opened up their new applications in many display devices. At present, there are different types of LEDs available on the market. LEDs are categorized on the basis of their characteristics, including colors/wavelength, the intensity of the emitted light, and so on. These different LED characteristics basically depend on important factors such as semiconductor materials; fabrication technology and encapsulation are also important determinants of LED characteristics. Usually, LEDs are manufactured from semiconductor materials such as GaAs, GaAsP, SiC, or GaInN. If mixed together in different ratios, these can produce a unique, distinctive wavelength of light. Currently, the production cost of blue and white LEDs is generally high due to the production costs of mixing two or more colors in the proper ratio inside the semiconductor materials.

4.1.3 Types of Light-Emitting Diode

LEDs can be broadly classified into two main categories:

- Visible LEDs
- Invisible LEDs

Visible LEDs are primarily used for switches and optical displays and for illumination purposes without the use of any photosensors. Invisible LEDs are used in applications including optical switches, analysis, and optical communications, and so on, with the use of photosensors.

4.1.4 Operation of LEDs

The LED is a p–n junction diode. It is a specially doped diode made up of a special type of semiconductor. When the light is emitted in the forward bias, it is called a *light-emitting diode*. The LED is a two-lead semiconductor light source. In 1962, Nick Holonyak, who was working for the General Electric Company, came up with the idea of the LED. Hence, the LED allows the flow of current in the forward direction and blocks the current in the reverse direction. The LED occupies a small area, less than 1 mm^2. LEDs are applied in various electrical and electronic projects. Figure 4.2 shows a schematic representation of the LED structure. When the diode is forward biased, the electrons and holes are moving fast across the junction and combining constantly, removing one another. Soon after an electron moves from the n-type to the p-type silicon, it combines with a hole, and then it disappears. Hence, it makes a complete atom, which is more stable, and it gives a little burst of energy in the form of a tiny packet or photon of light.

When the external supply voltage is applied to the free electrons, they gain a sufficient amount of energy to break the bonding with the parent atom. When a free electron leaves the parent atom, it leaves a vacant space in the valence shell. This vacant or empty space is called a *hole*. The valence electrons have almost the same energy levels. On the basis of the energy

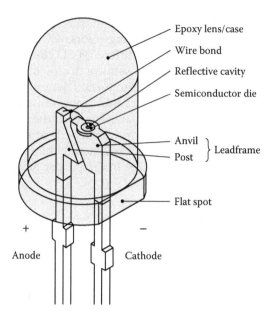

Epoxy lens/case

Wire bond

Reflective cavity

Semiconductor die

Anvil ⎫
          ⎬ Leadframe
Post   ⎭

Flat spot

+

−

Anode

Cathode

FIGURE 4.2
Internal structure of LED.

band diagram of a semiconductor, it consists of two bands, as shown in Figure 4.3. The lower and filled band, called the *valence band* (VB) consists of holes. The upper, empty band is called the *conduction band* (CB) and is filled with free electrons. The energy levels of free electrons in the CB must be higher than the energy levels of holes in the VB. During the electron–hole recombination process, free electrons in the CB lose some energy and recombine with the holes in the VB. The free electrons in the CB need not stay for a long time; after a short time, they move downward, lose their energy by emitting light of a certain wavelength, and finally recombine with the holes that are present in the VB. So, during each electron–hole recombination process, the charge carrier will emit some photon energy. The intensity of the emitted photon strictly depends on the energy gap (forbidden gap) between CB and VB. Hence, a semiconductor device with a large band gap emits high-intensity photons, whereas a semiconductor with a small band gap emits low-intensity photons. Thus, the output brightness of an LED depends on the materials and the forward current flowing through it. LEDs are always operated in forward bias due to the presence of the recombination process. However, it does not work in reverse bias, because minority charge carriers from the n-side and majority carriers from the p-side move away from the junction, so that the width of the depletion region increases and no recombination takes place; that is, no light is emitted from the device. Among the different semiconductor diodes, only LEDs can emit light in the visible region, because materials

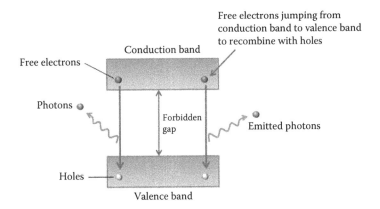

FIGURE 4.3
Energy band structure and process of light emission in LEDs.

used for the fabrication of LEDs are selected in such a way that the emission wavelength of the materials falls in the visible spectral region.

4.1.5 I-V Characteristics of LEDs

When the current flows through an LED it shows emission in a particular wavelength region. So, it needs a conventional current to flow across it; therefore, an LED is a current-dependent semiconductor device, and the intensity of the output light is directly proportional to the forward current passing through it. Usually, the LED is connected in a forward bias with respect to power supply, and a suitable resistor is connected in series to prevent the damage that can occur due to excess current flow through it. Figure 4.4 shows

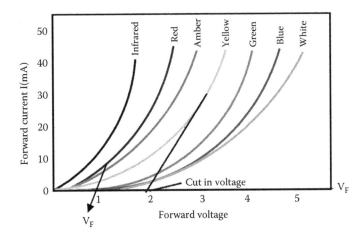

FIGURE 4.4
I-V characteristics of the different color LEDs.

the nature of the current–volt characteristics curve for different colors of LED. It is seen that with increasing the supply voltage current flow through the device also, increases. When the supplied voltage reaches the cut in voltage then the LED conducts and emits light, just like the forward-biased characteristics of the PN junction diode.

From this curve we can calculate the cut in voltage of the LED.

4.2 Semiconductor Materials for LEDs

The light is produced by the solid-state material; hence, LEDs are referred to as *solid-state devices*. Solid-state lighting consists of organic and inorganic LEDs and is superior to other sources of light such as incandescent and tungsten halogen lamps, gas discharge, fluorescent lamps, and soon. Semiconductor materials play an important role in light emission by supporting the active centers. The following subsections describe the main semiconductor compounds used in LEDs.

4.2.1 Indium Gallium Nitride

InGaN is a direct–band gap semiconductor material that exhibits strong band-to-band emission in the range of green to UV according to the indium content [16]. Fluctuations in indium content are thought to generate deep localized energy states, which are held to be responsible for the strong band-to-band emission at room temperature [17, 18]. Also, when indium is added to GaN, it significantly reduces nonradiative recombination centers, which arise due to certain point defects. Using this material as an active layer, Nakamura et al. successfully devised the first blue InGaN/AlGaN double heterostructure LED in 1994, and this was followed by the InGaN single quantum well structure LED in 1995. InGaN LEDs have recently been shown to have more than 50% external quantum efficiency in the deep blue range [19]. The most desirable advantage of the InGaN active layer is that it contains a large number of threading dislocations as a consequence of lattice mismatch [18]. The threading dislocations are thought to be an outcome of complex interactions between the interface energy and the nucleation density. However, a study of the literature shows that the photoluminescence (PL) intensity of an LED depends largely on the presence of localized energy states and has no association with the density of threading dislocations. In the case of red InGaN single quantum well LEDs, red emission originates due to carrier recombination taking place at the localized energy state because of the high In proportion. In the absence of indium, there will be no composition fluctuations that give rise to a strong localized state in the InGaN active layer. As a consequence, the quantum confined Stark effect, emerging from the piezoelectric field because of strain, becomes decisive and induces a spatial separation of electron and hole wave functions in the quantum well. Eventually, the wave function overlap decreases, and the interband recombination rate is reduced. This is how the efficiency of LEDs is scaled down in the absence or with a low amount of indium [20]. Thus, in conventional InGaN LEDs, variation in quantum well thickness and the relative proportion of indium plays a dominant role in adjusting the wavelength and enhancing the luminous brightness.

4.2.2 Aluminum Gallium Indium Phosphide

Aluminum gallium indium phosphide material is used for high-brightness LEDs emitting in the long-wavelength range of the visible spectrum. It is a low-cost source with high-power output and better color rendering capabilities. In the deep red emission region (around 650 nm), this semiconductor system is estimated to have near 100% internal quantum efficiency [21]. The addition of Al to the GaInP active region helps in achieving shorter emission wavelengths, including in the orange and yellow spectral range [22]. The AlGaInP system manifests low radiative efficiency at the direct–indirect crossover region due to direct–indirect transition of the band gap. However, the band gap energy at which direct–indirect transition occurs depends on the degree of randomness of the semiconductor material, and it is low for AlGaInP. As a consequence, AlGaInP systems can be used for various colors of the long-wavelength region of the visible range. By employing AlGaInP LEDs as a source of red LEDs, this allows the avoidance of the very large Stokes shift accompanying blue/red phosphor combinations [23].

4.2.3 Aluminum Gallium Arsenide

Aluminum gallium arsenide, which was developed in the late 1970s, was the first material system relevant for high-brightness LED applications. Aluminum gallium arsenide is basically a direct–band gap semiconductor and is lattice matched to GaAs substrates. It is best suited for visible spectrum LEDs that emit in the red wavelength region. The direct and indirect crossover in the band gap region occurs approximately at a wavelength of 621 nm, where the radiative efficiency of the AlGaAs system becomes extremely low due to direct–indirect transition [24]. To maintain high luminous efficiency, the emission energy must be lower than the band gap energy. AlGaAs LEDs have been developed as homostructures and single and double heterostructures. However, double heterostructures are the most efficient LEDs. The authenticity of AlGaAs systems is less than that of AlGaInP systems, because the high ratio of Al in AlGaAs active layers is more prone to corrosion and oxidation, thus lowering the lifetime and efficiency of the LED [25]. Researchers have reported deterioration in AlGaAs in the form of fissures and cracks when thick AlGaAs layers with nearly 85% Al content were exposed for a long time to ambient environmental conditions. Researchers have also found that thin AlGaAs layers are relatively more stable even for an Al content of 100% [26].

4.2.4 Gallium Phosphide

Gallium phosphide is an indirect–band gap semiconductor substrate that emits light in the green color wavelength. Since the conversion efficiency of GaP is 0.1%, it is doped with optically active impurities such as N, O, and Zn. These optically active impurities create an optically active level within the forbidden gap region of a semiconductor such that carriers recombine radiatively. The spectral transmission range of GaP is shifted to shorter wavelengths, owing to which it can be used for visible and near-infrared applications. GaP possesses superior mechanical durability, fracture toughness, and flexural strength. It is an appropriate material for optical devices that are exposed to harsh weather conditions and mechanical strain. Commercially, green LEDs with low brightness are fabricated using GaP doped with N. They are also used in indicator lamps [27].

4.3 Types of Phosphor

4.3.1 Blue Phosphor

Blue phosphor is a promising candidates for the development of WLEDs (White LEDs). Due to its small Stokes shift and short decay time, Eu^{2+} ion is the most commonly used blue-emitting activator. It exhibits broad near UV excitation and emission bands due to exposure of 5d electrons to the crystal environment. Examples of blue-emitting phosphors include $CaLaGa_3S_6O:Ce^{3+}$ [28], $LiSrPO_4:Eu^{2+}$ [29], $SrMg_2Al_{16}O_{27}:Eu^{2+}$ [30], $Ca_2PO_4Cl:Eu^{2+}$ [31], and so on. Recently, Gwak et al. have developed an efficient oxyhalide blue phosphor, $Sr_4OCl_6:Eu^{2+}$, with improved PL emission and conversion efficiency [32]. The oxyhalide phosphor Sr_4OCl_6 has demonstrated a broad excitation band ranging from 250 to 425 nm, which matches perfectly with the emission of near UV LED chips (λ_{max} = 395 nm). It emits intense blue light under λ_{ex} = 370 nm. Gwak et al. reported that the thermal stability of $Sr_4OCl_6:Eu^{2+}$ was much higher than that of the commercial blue phosphor, with only 9% emission loss for $Sr_4OCl_6:Eu^{2+}$ compared with 19% for commercial phosphor at 200°C. WLEDs fabricated with the synthesized $Sr_4OCl_6:Eu^{2+}$ on InGaN LEDs have shown ideal CIE chromaticity (0.32, 0.33) at the applied current of 20 mA, with the highest *color rendering index* (CRI) value of 82. Liu et al. have synthesized $NaSrBO_3:Ce^{3+}$, which shows high external quantum efficiency (89% of the blue-emitting commercial compound $BaMgAl_{10}O_{17}:Eu^{2+}$) [33]. Song et al.

have synthesized a novel blue-emitting $RbBaPO_4:Eu^{2+}$(RBP) phosphor via solid-state synthesis. The emission spectra of the RBP phosphors showed an intense purplish-blue-emitting band with a maximum at 430 nm. Based on the crystallographic information, it was found that RBP exhibited an orthorhombic structure. One phosphorus atom was coordinated with four oxygen atoms in a rigid tetrahedral structure, whereas only one Ba^{2+} site was observed in the RBP structure. According to experimental observations, the chromaticity of the RBP phosphors, along with their luminous efficacy, was found to be more stable than that of the commercially used $Sr_3MgSi_2O_8:Eu^{2+}$ phosphor [34]. Luminescence in rare earth–doped III–V group nitrides such as AlN, GaN, InGaN, and AlInGaN has also been intensively investigated because of their applications in blue–UV optoelectronic devices. There are two fundamental reasons for choosing nitrides for blue light sources. First is the large band gap, which encompasses the entire visible spectrum, and second is the strong chemical bonding, which makes the phosphor material very stable and resistant to degradation under conditions of intense light illumination.

4.3.2 Red Phosphors

Red-emitting phosphors with narrowband emission spectra play an essential role in the manufacture of high-quality LEDs. Silicon-based nitride compounds are considered as good host lattices for red luminescent materials [35, 36]. The rare earth activators that are capable of contributing red emission in various nitride, borate, phosphate, sulfide, silicate, vanadate, and tungstate hosts are Eu^{3+}, Pr^{3+}, Sm^{3+}, and Eu^{2+} [37]. The $^7F_0$-$^5L_6$ and $^7F_0$-$^5D_2$ transition wavelengths of Eu^{3+} are a reasonable match to the emission wavelengths of near UV InGaN and GaN semiconductor chips, respectively. Also, the site symmetry of Eu^{3+} provides an opportunity to tune luminescent emission from orange to red by crystal engineering. Tian et al. have reported luminescent characteristics of flower-like-shaped red phosphor $(Y_2MoO_4)_3:Eu^{3+}$ synthesized through a simple co-precipitation method. $Y_2MoO_4)_3:Eu^{3+}$ phosphor showed higher color saturation and better chromaticity coordinates than commercially available $Y_2O_2S:Eu^{3+}$ red phosphor [38]. Uheda et al. have developed red phosphor $CaAlSiN_3:Eu^{2+}$ with quantum output seven times higher than conventional red phosphor $La_2O_2S:Eu^{3+}$ under excitation of 405 nm. Also, $CaAlSiN_3:Eu^{2+}$ shows much less thermal quenching, ascribed to the rigid tetrahedral network of $[SiN_4]$ and $[AlN_4]$ in its crystal structure, leading to a small Stokes shift and higher quantum output [39]. Manipulating the coordination of the Mn^{2+} ion by mixing different anions is also one suggested way to increase opportunities for designing novel Mn^{2+}-doped phosphors with unconventional properties for WLED applications. Recently, Tang et al. reported a series of $(K_xNa_{1-x})SiF_6:Mn^{2+}$ red phosphors with a sequence of continuous structural phase transformations [40]. After the study of different phases, including trigonal, orthorhombic, and cubic, it was found that the emission intensity of cubic phosphor (K_xNa_{1-x}) SiF_6:Mn^{2+} was increased more than five times with respect to that of trigonal (K_xNa_{1-x}) SiF_6:Mn^{2+}. This study evidently provided an unambiguous demonstration of the decisive relationship between the structure and optical properties of matter [41]. Similarly, Munirathnappa et al. made an attempt for the first time to tune the band gap of $NaEu(WO_4)_2$ (NaEuW) phosphor to enhance its red emission via a different dose of high-energy electron beam (EB) irradiation. The study reveals the influence of EB irradiation on the crystal structure, morphology, and red fluorescence properties of NaEuW to generate efficient white light [42].

4.3.3 Green Phosphors

To achieve standard white emission, the emission intensity of green- and red-emitting phosphors must be strong. Tb^{3+} is prominently used as an activator for green-emitting phosphors. Tb^{3+}-activated phosphors are found to be highly efficient for excitation in the near UV range [43]. Eu^{2+} and Ce^{3+} also yield green luminescence. Eu^{2+} when used as a sensitizer for energy transfer with Tb^{3+} and Ce^{3+} as an activator is responsible for the enhancement of green emission intensity. A green color–emitting phosphor often has an emission peak at 510–550 nm. Nazarov et al. have successfully synthesized green phosphors based on strontium thiogallate $(SrGa_2S_4 + MgGa_2O_4)$:Eu^{2+} by employing a solid-state reaction [44]. The $SrGa_2S_4$ thiogallate compounds belong to the orthorhombic crystal class with space group D_{2h}^{24} [45], whereas $MgGa_2O_4$ belongs to the cubic structure class with space group O_4^7. Experimental results have clearly demonstrated [45, 46] 15% higher PL emission for $(SrGa_2S_4 + MgGa_2O_4)$ than for the best commercially available green emitter, SrG_2S_4. Kijima et al. have developed a Ce^{3+}-activated $CaSc_2O_4$, which exhibited green luminescence with a peak wavelength of 515 nm under blue light excitation [46]. The host phosphor showed an orthorhombic $CaSc_2O_4$ structure with the Ce ion existing in an eight-coordinated Ca position. The luminescence efficiency of the reported oxide crystal was found to be better than that of commercially available $Y_3Al_5O_{12}$:Ce^{3+} phosphor. Zhang et al. have synthesized Eu^{2+}-doped $BaSi_2O_5$ as an efficient phosphor for LEDs [47]. In this phosphor, the green emission band was centered on 500 nm. The *photoluminescence excitation* spectrum revealed a broad band from 260 to 400 nm with strong intensity around 340–370 nm due to crystal field splitting and transition between ground state and $4f^7$. The phosphor displayed high current saturation and high resistance under low-voltage electron bombardment. Hirosaki et al. have synthesized a green oxynitride phosphor doped with Eu^{3+} and β-SiAlON. The β-SiAlON: Eu^{2+} phosphor generates intense green emission with the peak maximum located at 538 nm [48]. The broad excitation range from 300 to 420 nm enables the β-sialon:Eu^{2+} phosphor to emit strongly under near UV (400–420 nm) or blue (420–470 nm) light excitation. $Sr_6BP_5O_{20}$:Eu^{2+} [49],

$Ca_3Y_2(Si_3O_9)_2$:Ce^{3+}, Tb^{3+} [50], $SrAl2O_4$:Eu^{2+} [51], and $Ca_2MgSi_2O_7$:Eu^{2+}, Mn^{2+} [52] are a few more examples of green phosphors.

4.3.4 Yellow Phosphors

Yellow phosphors with garnet structure and the general formula $A_3B_2C_3O_{12}$ doped with rare earth ions have become extremely popular for a broad range of applications. The first WLEDs were based on InGaN blue LEDs and yellow yttrium aluminum garnet doped with Ce as phosphor. YAG:Ce phosphors demonstrate superior optical characteristics. However, because of high thermal quenching, changes occur in the chromaticity of WLEDs when used. Hence, there was a strong incentive to develop novel yellow phosphors that undergo low thermal quenching and create warm white light. Seto et al. have investigated a novel $La_3Si_6N_{11}$:Ce yellow phosphor that showcases low thermal quenching [53]. Oxynitride and nitride hosts activated by Ce^{3+} possess large crystal field splitting. The low centroid level of 5d states of Ce^{3+}, due to coordination of N^{3-} with high covalency, provides a decrease in energy difference between the 4f and 5d states. Consequently, nitride phosphors tend to have long wavelengths suitable for WLEDs. Pan et al. have introduced a new yellow $Ba_{0.93}Eu_{0.07}Al_2O_4$ phosphor for WLEDs through single emitting center conversion [54]. It has an orthorhombic lattice structure and highlights a broad yellow PL band with sufficient red spectral component. $Ba_{0.93}Eu_{0.07}Al_2O_4$ phosphor produces warm white light emission with correlated color temperature < 4000 K and CRI > 80 when combined with a blue LED (440–470 nm). The warm white light with better CRI was attributed to the presence of sufficient red component in the broad yellow emission band of $Ba_{0.93}Eu_{0.07}Al_2O_4$. Jang et al. have reported a yellow-emitting Sr_3SiO_5:Ce^{3+}, Li^+ phosphor [55]. It showed high luminous efficiency under near UV light as well as under blue light. When Sr_3SiO_5:Ce^{3+}, Li^+ was coated on a 380 nm GaN-based near UV LED, it showed a bright yellow light. Therefore, it is supposed that Sr_3SiO_5:Ce^{3+}, Li^+ phosphor will exhibit excellent white light properties if coated on a near UV LED together with red, blue, and green phosphors, because deficient regions in the visible spectrum will be compensated by the broad yellow emission from this phosphor. J. Zhang et al. have proposed a yellow-emitting nitride phosphor $SrAlSi_4N_3$:Ce^{3+}, which has low thermal quenching and a broad emission band centered at 555 nm with a bandwidth of 115 nm. This particular phosphor belongs to an orthorhombic crystal system and contains spatial infinite chains of edge-sharing $[AlN_4]$ tetrahedra running along [001] in the lattice [56]. Two Al sites were positioned in $[AlN_4]$ chains, and both of them were coordinated with N_1, N_2, N_4, and N_8 atoms. The electrostatic valence sum of N_1 was found to be 2.5, which caused 17% divergence from the valence of the N atom. This divergence value exceeded the critical value and thus kept the compound stable. Ca-α-SiAlON: Eu^{2+}, $Zn_3V_2O_8$:Bi^{3+}, and $NaBaBO_3$:Sm^{3+} are a few more good examples of yellow-emitting phosphors.

Ultimately, this chapter is a brief summary of LED basics and different phosphors used for efficient LED lighting. Though there has been tremendous progress, there are still some unresolved challenges to be solved so as to discover LEDs with maximizing conversion efficiencies and long lifetime.

4.4 Fabrication of White LEDs (WLEDs)

To fulfill the demand for daylight, WLEDs have been extensively investigated. Various improvements have been made in the efficiency and stability of WLEDs. There are two primary mechanisms that are being used to produce white light, as shown in Figure 4.5. The different methods of mixing colors to produce white light from an LED [16], are discussed in the following:

1. Blue LED + Green LED + Red LED = White LEDs
2. Near UV or UV LED + Yellow phosphor/RGB phosphor = White LEDs

In the first method (the RGB method), red-, green-. and blue-emitting individual LEDs are used to produce a nearly white light emission, as shown in Figure 4.5. In the second method, the red, green, and blue color–emitting materials are blended into a single LED to produce white light emission. Also, a blue or UV LED can be used to excite a phosphor material to produce white light (see Figure 4.5) in a process similar to that in fluorescent lamps. Each of these methods has both advantages and disadvantages; therefore,

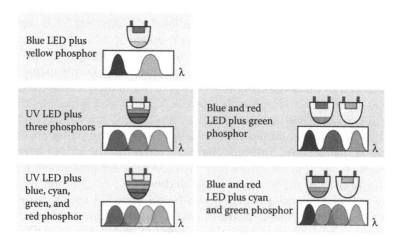

FIGURE 4.5
Methods of white light generation. (From www.lightemittingdiodes.org.)

the "whiteness" of the light produced is essentially engineered to suit the human eye. The second method is much more efficient than the first method and more commonly used in many lighting devices to produce white light.

4.4.1 Method 1: RGB Method

White light can be generated by mixing the three primary colors of light (red, green, and blue), as shown in Figure 4.6. Hence, this method is called *multicolor white LEDs* (or *RGB LEDs*). This is the most common approach for producing white light [17]. The white light emission is controlled by the electronic circuit to blend and diffuse the different colors; the individual LEDs can have different emission pattern and emission intensity, and this can lead to a variation in white light emission. This is the most popular method used in many applications because of the ease of mixing different colors, and LEDs fabricated through this mechanism have higher quantum efficiency in producing white light. The RGB method is an additive color method, in contrast to the subtractive color system of pigments, dyes, and inks, which present color to the eye by reflection rather than emission. For example, the green color in the subtractive color system can be produced by mixing yellow and blue colors, whereas in the additive color system, the mixing of red and green colors gives yellow color, and green color cannot be produced in simple combination. The additive color is not a property of light; it is a result of the way the eye detects the color. There is a significant difference between yellow light (580 nm) and a yellow light obtained through combining red and green light. Both these colors look similar to the eye, so we do not recognize the difference. Today different types of multicolor (di-, tri-, and tetrachromatic) WLEDs are available in markets. The key factors that are to be optimized by these different approaches include color stability, CRI, and efficiency. Color rendering (CR) is the ability of a light source to reproduce colors faithfully in comparison with an ideal or natural light source [18]. Very often, LEDs with higher efficiency have poor CR. For example, two-color (dichromatic) WLEDs have luminous efficiency (120 lm/W) but possess lower CR ability. However, tetrachromatic white LEDs have good CR ability, but they often have poor luminous efficiency. For both good luminous efficiency and fair

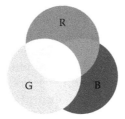

FIGURE 4.6
White color production through a mixture of red, green, and blue colors.

CR ability, trichromatic WLEDs are used. Multicolor LEDs are used not only in white light production but also in generating different colored light; in this way, light of different colors spanning the entire visible spectrum can be produced. Apart from this benefit, the light (emission power) produced by this method decays exponentially with rising temperature, suggesting poor color stability. This type of problem can hinder the applicability of such a method in large-scale production. Research is ongoing to overcome this issue. Moreover, multicolor LEDs can never deliver good-quality lighting, because LED is a narrowband source of light; this opens up the use of LEDs in displays, either as the backlight of a liquid crystal display or directly as LED-based pixels.

4.4.2 Method 2: UV LED + Yellow Phosphor/RGB Phosphor

RGB phosphor–based LEDs involve a coating of phosphors with different colors on a single (UV or blue) colored LED (made of InGaN) to generate white light, as shown in Figure 4.6. LEDs of this type are commonly known as *phosphor-based* or *phosphor-converted white LEDs* (pcLEDs). This technique employs a blue or UV LED coated with a yellow-emitting phosphor. The shorter-wavelength photons from the LED chip excite the phosphor, resulting in the generation of yellow photons in the phosphor. The shorter-wavelength photons (blue) and longer-wavelength photons (yellow) combine to generate a white light [19]. The UV/blue light from the LEDs undergoes a Stokes shift and is transformed to a longer-wavelength emission. Phosphors of different emission colors can be coated on LEDs, and a multicolor or white light emission can be achieved. LEDs fabricated by this technique possess improved CR ability but have two major concerns. The first is heat loss due to the Stokes-shifted emission, and the second is the stability of the particular phosphor used in the LEDs. The light emission efficiency depends on the spectral distribution of the resultant light output and the original wavelength of the LED. For example, the efficiency of cerium-doped yttrium aluminum garnet (YAG—a yellow-emitting phosphor)–based white LEDs ranges from three to five times that of the original blue LED because of the human eye's greater sensitivity toward yellow than blue. The simplicity of manufacturing high-intensity WLEDs makes this method more popular than Method 1. The design and production of a light source using phosphor conversion is easy and cheaper than the complicated RGB method. The majority of the WLEDs presently in the market are manufactured using this technique (Figures 4.7 and 4.8).

One of the major challenges with WLEDs is to improve their efficiency, and this can be done by employing more efficient phosphors in LEDs. To date, the most efficient yellow-emitting phosphor is the YAG phosphor with less than 10% Stokes shift loss. Self-absorption in LED chip and packing materials can further contribute another 10%–30% efficiency loss. Efforts are ongoing to improve the light emission efficiency as well as to increase the higher operating temperature. One way to improve the overall efficiency of

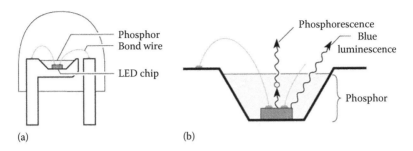

FIGURE 4.7
(a) Structure of white LED consisting of a GaInN blue LED chip and a phosphor encapsulating the die. (b) Wavelength converting phosphorescence and blue luminescence.

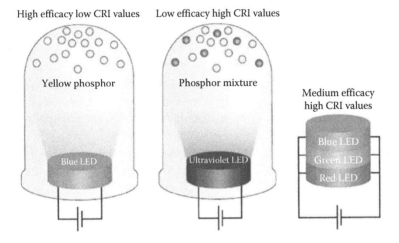

FIGURE 4.8
Three types of white LEDs for lighting: (a) blue LED plus yellow phosphors, (b) ultraviolet LED plus three phosphors, (c) three LEDs: red, green, blue connected in parallel. (From S. Pimputkar, J. S. Speck, S. P. DenBaars, S. Nakamura, *Nat. Photon.* 3, 180–182, 2009.)

LEDs is by using more suitable and efficient phosphor materials in LEDs. Often, white light–emitting phosphors offer better CR ability than the white light obtained through RGB, with improved light emission efficiency. Due to its superior CR and improved efficiency, phosphor white is the most feasible white light–producing technique. The white light emission in the phosphor white is determined by the dominant blue emission from the LED and the composition of the phosphor. The phosphor coating thickness determines the color temperature of the LED. During manufacture of the LED, the variation in color is optimized by controlling the thickness and composition of the phosphor layer. An alternative way to produce a white light is to make use of a coating of a mixture of high-efficiency europium-based phosphors that emit red and blue on the UV LED. In addition to this, the copper and aluminum-doped zinc sulfide ($ZnS:Cu$, Al) phosphor, which emits green light, can be employed for white light production. This approach is similar to that of fluorescent lamps. This approach gives rise to more efficient white light emission, relatively higher than in the case of YAG:Ce phosphor coated on blue LEDs, which has a larger Stokes shift, resulting in more nonradiative losses but better CR ability.

4.5 Special Features of LEDs

LEDs are solid-state devices. LEDs offer several advantages over the conventional fluorescent lamps:

1. *Light generated by LEDs is directional*: LEDs produce all-forward directional lighting, not omnidirectional as in the case of the conventional light bulb. In general, beam angle is around 140°.

2. *LEDs can generate different light colors*: LEDs can be fabricated for multicolor light emission, which is very difficult in conventional fluorescent lamps. A mixture of RGB light or different phosphors can generate white light.

3. *Temperature will affect LED efficiency*: LEDs themselves generate heat, which can affect efficiency as well as LED lifespan. In general, a 10° increase in temperature reduces lumen output by 5%–7%. Maintaining the p–n junction temperature under 75° can enable LEDs to last for over 50,000 hours.

4. *Low energy consumption*: 100 lm/W in commercial applications, while over 200 lm/W is achieved in laboratories. A reduction to 1/5–1/10 of the power consumption of conventional lighting is achievable; in other words, an energy saving of 80%–90%.

5. *Long life*: No fragile parts to be broken, as a conventional light bulb. The lumen output can decay, but the light will rarely burn out or die. A well-designed luminaire can be expected to achieve over 70% lumen maintenance after 50,000 hours' use.

4.5.1 Challenges

4.5.1.1 Phosphor with Good CRI

Recently, researchers and industries have been making efforts to improve the quality of white light generated by LEDs. Thus, this is the main challenge in LED research. LEDs have so many advantages compared with fluorescent lamps in terms of high efficiency, better thermal stability, power saving and environment-friendly characteristics, long lifespan, and so on [57–59]. But the currently used commercial phosphor (Red, Green, and Blue) has some drawbacks during the production of white light, because it contains some red component during the light emission. This is we cannot get high-brightness white light. Researchers proposed another method to overcome this deficiency and obtain intense white light emission with a suitable combination of a blue-emitting GaN LED chip coupled with yellow-emitting phosphors (YAG:Ce^{3+}) [60–63], but the quality of light obtained with this method also faces some drawbacks, such as poor color rendering index (CRI = 70–80) and high color temperature (CCT~7750 K), which restrict the practical applications of LEDs in indoor lighting [64]. Another method used to eliminate the above-mentioned issues is to combine an InGaN-based LED chip with commercially used red, green, and blue phosphors. But it requires more efficient red, green, or blue phosphors, and there are only a few phosphors available for such applications that can be effectively excited by UV light to improve the luminous efficiency and CRI of devices. Thus, for the development of such energy-efficient phosphors, the selection of host materials plays an important role. Therefore, we require ideal phosphors to fabricate WLEDs, having excellent light-emitting materials with CIE coordinates (0.33, 0.33), the corresponding CCT between 2500 and 6500 K, and better CRI, approximately equal to 80 [65, 66].

4.5.1.2 Search for Novel Red Phosphors

For the past few decades, Y_2O_3:Eu^{3+} has acted as a promising red-emitting phosphor, which is commercially used in many light devices, but they have faced some drawbacks due to the lack of strong absorption bands in the blue spectral region. So, there is need to develop novel luminescent materials that show an intense red emission under the excitation of blue-emitting diodes, which is very interesting because of their high eye sensitivity, high CRIs, and low CCTs. They would also show the great benefit of overcoming the drawbacks of existing WLEDs.

4.5.1.3 Luminous Flux

Luminous flux is simply called *luminous power.* It is the measure of the perceived power of light, and it differs from the radiant flux, that is, the measure of the total power of electromagnetic radiation within the wavelength range from 390 to 770 nm. The luminous flux is adjusted to reflect the varying sensitivity of the human eye to different wavelengths of the visible spectral region. However, the existing LEDs lack luminous flux. Thus, in LED research, more weighting is being given to the increase of luminous flux, which can be achieved only by increasing the area of the semiconductor chips and operating supply voltage. This will produce higher junction temperature and strong temperature gradients, which can degrade the optical performance and reduce the reliability of the device. Thus, to avoid such issues, further improvements in device structure and materials with excellent properties are needed, with some advanced characterization techniques to study the important properties LEDs.

4.5.1.4 Thermal Stability

Currently, the demand for high-power LEDs is increasing due to their high energy efficiency and longer lifespan and their vast applications in display technology; also, they are used for outdoor lighting purposes. But the lifetime of the devices in such applications seems to be short [67]. However, the quality of light also changes when the device is in continuous operating mode due to the occurrence of thermal degradation mechanisms in LED materials at high temperature. There are now many reports available on LED degradation at high temperature, and some standard test measurements are also available, such as LM-80 and TM-21, to evaluate the thermal stability of these high-power LEDs [68–70]. Thus, to improve the quality of solid-state lighting (SSL) devices, there is a need to develop novel, highly efficient and thermally stable materials for LED devices.

4.5.1.5 Environmental and Humidity Effects

The lifetime of LEDs can be studied with the help of accelerated stress levels and the estimated lifetimes of the device under normal operating conditions. Tran et al. [69] reported that at constant humidity with varying temperature, the degradation mechanisms vary widely under different test conditions. They showed that the degradation of LEDs followed three steps, unlike integrated circuits, which show a simple degradation graph. Generally, there are many challenges still remaining to study humidity-induced degradation in LEDs, and some of them are listed here.

- As discussed earlier, reliability tests for LED devices can be performed in accelerated conditions, and thermal degradation can vary between accelerated and normal operating conditions, so that some standard test methods do not work for this purpose. Therefore, a new method is required to solve this difficulty. Finite element modeling must be used to determine the materials and device parameters in accelerated stress conditions and then apply these values to calculate the lifespan of LEDs in normal conditions.
- However, during the finite element modeling, the existing infinite element modeling for higher temperature and humidity is also a challenging task. Thus, further innovation is still needed in this area.
- Thus, regarding the degradation mechanisms at various conditions, the quality of LEDs should depend on environmental conditions. But in this regard, no standard test procedure can be established to study all parameters. Thus, the design of appropriate test procedures will be the next challenge in this field.

- To perform the finite element model, material parameters are required. But due to commercial storage conditions, such parameters are not usually available, and this will result in modeling difficulties, which are also challenging.

4.5.2 Bioluminescent Materials for LEDs: A New Approach

Biologically derived fluorescent proteins are an attractive alternative to current color-conversion materials for solid-state lighting applications, such as inorganic phosphors, organic dyes, and nanocrystal quantum dots [71]. Scientists from Germany and Spain have discovered a way to create a Bio LED by packaging luminescent proteins in the form of rubber. This innovative device gives off a white light, which is created by equal parts of blue, green, and red rubber layers covering one LED, thus rendering the same effect as with traditional inorganic LEDs but at a lower cost. Increasingly popular, LEDs are the light of choice for the European Union and the United States when it comes to creating lighting devices of the future. This preference can be attributed to the fact that LEDs are more efficient than traditional incandescent bulbs and more stable than energy-efficient light bulbs. Despite their advantages, however, LEDs are manufactured using inorganic materials that are in short supply—such as cerium and yttrium—meaning that they are more expensive and difficult to sustain in the long run. Additionally, WLEDs produce a color that is not optimal for eyesight, since they lack a red component, which can psychologically affect individuals exposed to them for long periods of time. Now, however, a German-Spanish team of scientists has drawn inspiration from nature's biomolecules in search of a solution. Their technique consists of introducing luminescent proteins into a polymer matrix to produce luminescent rubber. This technique involves a new way of packaging proteins, which could end up substituting the technique used to create LEDs today [72]. They developed a technology and a hybrid device called *Bio LED* that uses luminescent proteins to convert the blue light emitted by a 'normal' LED into pure white light. It is always necessary to have either a blue or an ultraviolet LED to excite the rubbers that are put over the LED to make it white. In other words, we can combine blue LED/green rubber/red rubber or ultraviolet LED/blue rubber/green rubber/red rubber. The result is the first Bio LED that gives off a pure white light created by similar proportions of the colors blue, green, and red, all while maintaining the efficiency offered by inorganic LEDs. Blue or ultraviolet LEDs are much cheaper than white ones, which are made of an expensive and scarce material known as *YAG:Ce* (cerium-doped yttrium aluminum garnet). The idea is to replace it by proteins. Bio LEDs are simple to manufacture, and their materials are of low cost and biodegradable, meaning that they can easily be recycled and replaced. The high stability of these proteins is evident from the fact that the luminescent properties remain intact during months of storage under different environmental conditions of light, temperature, and humidity. Scientists are already working on optimizing this new elastic material to achieve greater thermal stability and an even longer operating lifetime. They are addressing how to optimize the chemical composition of the polymer matrix in addition to using proteins that are increasingly more resistant to device operating conditions. The goal is to make this new Bio LED more accessible on an industrial scale in the not too distant future.

Initially, trees with naturally glowing leaves were assumed to be an eco-friendly alternative to street lamps. Researchers from the Academia Sinica and the National Cheng Kung University in Taipei and Tainan have implanted glowing, sea urchin–shaped gold nanoparticles, known as *bio light-emitting diodes* or Bio LEDs, inside the leaves of a plant. Luciferases are enzymes that catalyze reactions in living organisms that result in light emission. There are many different luciferases; different in the sense that the genes and proteins involved are unrelated in evolution and evidently originated and evolved independently [73].

The new nanoparticles could replace electricity-powered street lights with a biologically powered light that removes CO_2 from the atmosphere 24 hours a day. In the future, Bio LEDs could be used to make roadside trees luminescent at night. This will save energy and absorb CO_2, as the Bio LED luminescence will cause the chloroplasts to conduct photosynthesis. The gold, sea urchin–shaped nanoparticles are the key to turning a material that normally absorbs light into one that emits it [74]. Chlorophyll, the photosynthetic pigment that gives leaves their characteristic green color, is widely known for its ability to absorb certain wavelengths of light. However, under certain circumstances, such as being exposed to violet light, chlorophyll can also produce a light of its own. When exposed to light with a wavelength of about 400 nm, the normally green-colored chlorophyll glows red. Violet light is hard to come by, though, especially at night, when glowing leaves would be useful to drivers and pedestrians. The scientists needed a source of violet light and found it in the gold nanoparticles. When shorter wavelengths of light, invisible to the human eye, hit the gold nanoparticles, they get excited and start to glow violet. That violet light strikes the nearby chlorophyll molecules, exciting them, and the chlorophyll then produces red light. The scientists hope that trees treated with the gold nanoparticles would produce enough light that they could replace electric or gas street lights.

4.5.3 Advantages of LEDs

- *Energy benefits*: Currently, 22% of electricity use is for indoor and outdoor lighting applications. Therefore, LED-based lighting is more efficient and has energy-saving potential compared with incandescent and fluorescent lighting devices.

- *Environmental benefits*: LED-based devices save energy and produce less heat during operation as compared with gas discharge lamps and tungsten filament lamps, and one more advantage of this device is that it does not contain any toxic gas (e.g., mercury). The use of such devices may help to reduce the number of coal energy generation plants, so they are beneficial for the environment.

- *Economic benefits*: At least 10% energy saving is equivalent to financial savings of a billion dollars per year.

References

1. P. E. Dodds, I. Staffell, A. D. Hawkes, F. Li, P. Grünewald, W. McDowall, P. Ekins, *International Journal of Hydrogen Energy* 40 (2015) 2065.
2. M. Reza Maghami, S. N. Asl, M. E Rezadad, N. A. Ebrahim, C. Gomes, *Scientometrics* 105 (2015) 759.
3. M. S. Dresselhaus, I. L. Thomas, Alternative energy technologies, *Nature* 414 (2001) 332–337.
4. A. Franco, M. Shaker, D. Kalubi, S. Hostettler, A review of sustainable energy access and technologies for healthcare facilities in the Global South, Sustainable Energy Technologies and Assessments, 22 (2017) 92.
5. J. W. Sun, Changes in energy consumption and energy intensity: A complete decomposition model, *Energy Economics* 20 (1998) 85.
6. E. Fred Schubert, T. Gessmann, J. K. Kim, Light emitting diodes, *Encyclopedia of Chemical Technology* (2005) 1. doi.org/10.1002/0471238961.1209070811091908.a01. pub2.
7. M. R. Krames, O. B. Shchekin, R. Mueller-Mach, G. O. Mueller, L. Zhou, G. Harbers, M. G. Craford, Status and future of high-power light-emitting diodes for solid-state lighting, Journal of Display Technology, 3 (2007) 160.
8. H. Amano, T. Asahi, I. Akasaki, *Japanese J. Applied Physics* 29 (1990) L205.
9. Y. Nanishi, *Nature Photonics* 8 (2014) 884.
10. J. Zhou, W. Yan, Power Electronics Specialists Conference (2007), Orlando, FL, 436.
11. T. Komine, M. Nakagawa, *IEEE Transactions on Consumer Electronics* 50 (2004) 100.
12. E. H. Perea, *Appl. Phys. Lett.* 36 (1980) 978.
13. K. Konnerth, C. Lanza, Delay between current pulse and light emission of a gallium arsenide injection laser, *Appl. Phys. Lett.* 4 (1964) 120.
14. M. R. Lorenz, G. D. Pettit, R. C. Taylor, Band gap of gallium phosphide from 0 to 900 K and light emission from diodes at high temperatures, *Phys. Rev.* 171 (1968) 876.
15. S. Ogawa, M. Imada, S. Yoshimoto, M. Okano, S. Noda, *Science* 305 (2004) 227.
16. S. Nakamura, T. Mukai, *Jpn. J. Appl. Phys.* 31 (1992) L1457.
17. S. Chichibu, T. Azuhata, T. Sota, S. Nakamura, *Appl. Phys. Lett.* 70 (1997) 2822.
18. Y. Narukawa, Y. Kawakami, Sz. Fujita, Sg. Fujita, S. Nakamura, *Phys. Rev. B* 55 (1997) 1938R.

19. S. D. Lester, F. A. Ponce, M. G. Crawford, D. A. Steigerwald, *Appl. Phys. Lett.* 66 (1995) 1249.
20. M. R. Krames, O. B. Shchekin, R. Mueller-Mach, G. O. Mueller, L. Zhou, G. Harbers, M. G. Craford, *J. Display Technology* 3 (2007) 160.
21. T. Mukai, M. Yamada, S. Nakamura, *Jpn. J. Appl. Phys.* 38 (1999) 3976.
22. J. M. Phillips, M. E. Coltrin, M. H. Crawford, A. J. Fischer, M. R. Krames, R. Mueller-Mach, G. O. Mueller, et al., *Laser & Photon. Rev.* 4 (2007) 307.
23. M. Ikeda, K. Nakano, Y. Mori, K. Kaneko, N. Watanabe, *J. Cryst. Growth* 77 (1986) 380.
24. T. Gessmann and E. F. Schubert, Journal of Applied Physics **95**, (2004) 2203.
25. H. Ishiguro, K. Sawa, S. Nagao, H. Yamanaka, S. Koike, *Appl. Phys. Lett.* 43 (1983) 1034.
26. S. Ishimatsu, Y. Okuno, *Dev. Technol.* 4 (1989) 21.
27. F. A. Kish, R. M. Fletcher, AlGaInP light-emitting diodes in *High Brightness Light Emitting Diodes* edited by G. B. Stringfellow and M. G. Craford, Semiconductors and Semimetals 48, Academic Press, San Diego (1997).
28. R. J. Yu, J. Wang, M. Zhang, J. H. Zhang, H. B. Yuan, Q. Su, *Chem. Phys. Lett.* 453 (2008) 197.
29. Z. C. Wu, J. Shi, J. Wang, M. L. Gong, Q. Su, *J. Solid State Chem.* 179 (2006) 2356.
30. V. B. Pawade, S. J. Dhoble, *Luminescence* 26 (2011) 722.
31. R. Yu, C. Guo, T. Li, Y. Xu, *J. Phys. Chem. C* 118 (2014) 2686.
32. S. J.Gwak, P. Arunkumar, W.B. Im, *J. Phys. Chem. C* 118 (2014) 2686..
33. Wei-Ren Liu, Chien-Hao Huang, Chih-Pin Wu, Yi-Chen Chiu, Yao-Tsung Yeh, Teng-Ming Chen, *J. Mater. Chem.* 21 (2011) 6869.
34. Hee Jo Song, Dong Kyun Yim, Hee-Suk Roh, In Sun Cho, Seung-Joo Kim, Yun-Ho Jin, Hyun-Woo Shim, et al., *J. Mater. Chem. C* 1 (2013) 500.
35. Y. Q. Li, A. C. A. Delsing, G. de With, H. T. Hintzen, *Chem. Mater.* 15 (2005) 4492.
36. K. Uheda, N. Hirosaki, Y. Yamamoto, A. Naito, T. Nakajima, H. Yamamoto, *Electrochem. Solid State Lett.* 9 (2006) H22.
37. R. Le Toquin, A. K. Cheetham, *Chem. Phys. Lett.* 423 (2006) 352.
38. Y. Tian, X. Qi, X. Wu, R. Hua, *B. Chen J. Phys. Chem. C* 113 (2009) 10767.
39. K. Uheda, N. Hirosaki, Y. Yamamoto, A. Naito, T. Nakajima, H. Yamamoto, *Electrochem. Solid State Lett.* 9 (2006) H22.
40. F. Tang, Z. C. Su, H. Ye, S. J. Xu, G. Wang, Y. Cao, W. Gao, et al., *J. Mater. Chem. C* 5 (2017) 12105. DOI:10.1039/C7TC04695B.
41. R. Kasa, S. Adachi, *J. Electrochem. Soc.* 159 (2012) J89.
42. A. K. M, N. Sundaram, J. Dwivedi, V. C. Petwal, *New J. Chem.* (2018). DOI:10.1039/C7NJ04094F.
43. R. C. Ropp, *The Chemistry of Artificial Lighting Devices.* Amsterdam: Elsevier (1993) 486.
44. Mihail Nazarov, Do Young Noh, Clare Chisu Byeon, Hyojung Kim, *J. Appl. Phys.* 105 (2009) 073518.
45. B. Eisenmann, M. Jakowski, W. Klee, H. Schäfer, *Rev. Chim. Miner.* 20 (1983) 255.
46. Yasuo Shimomura, Tomoyuki Kurushima, Naoto Kijima, *J. Electrochem. Soc.* 154 (2007) J234.
47. Qiang Zhang, Qian Wang, Xicheng Wang, Xin Ding, Yuhua Wang, *New J. Chem.* 40 (2016) 8549–8555.

48. N. Hirosaki, R.-J. Xie, K. Kimoto, T. Sekiguchi, Y. Yamamoto, T. Suehiro, M. Mitomo, *Appl. Phys. Lett.* 86 (2005) 211905.

49. M. Zhang, J. Wang, W. Ding, Q. Zhang, Q. Su, *Appl. Phys. B* 86 (2007) 647.

50. Y. C. Chiu, W. R. Liu, Y. T. Yeh, S. M. Jang, T. M. Chen, *J. Electrochem. Soc.* 156 (2009) 221.

51. K. Y. Jung, H. W. Lee, H. K. Jung, *Chem. Mater.* 18 (2006) 2249.

52. C. K. Chang, T. M. Chen, *Appl. Phys. Lett.* 90 (2007) 161901.

53. T. Seto, N. Kijima, N. Hirosaki, *ECS Trans.* 25 (2009) 247.

54. X. Li, J. D. Budai, F. Liu, J. Y Howe, J. Zhang, X.-J. Wang, Z. Gu, et al., *Light Sci. Appl.* 2 (2013) 50.

55. H. S. Jang, D. Y. Jeon, *Appl. Phys. Lett.* 90 (2007) 041906.

56. C. Hecht, F. Stadler, P. J. S. Schmidt, J. S. Günne, V. Baumann, W. Schnick, *Chem. Mater.* 21 (2009) 1595.

57. J. Li, J. Yan, D. Wen, W. U. Khan, J. Shi, M. Wu, Q. Su, P. A. Tanner, Advanced red phosphors for whitelight-emitting diodes. *J. Mater. Chem. C* 4 (2016) 8611.

58. P. Du, L. Luo, X. Huang, J. S. Yu, *J. Colloid Interface Sci.* 514 (2018) 172.

59. X. Huang, *J. Alloys Compd.* 628 (2015) 240.

60. B. Li, X. Huang, *Ceram. Int.* (https://doi.org/10.1016/j. ceramint.2017.12.082.)

61. X. Huang, H. Guo, B. Li, *J. Alloys Compd.* 720 (2017) 29.

62. M. Peng, X. Yin, P. A. Tanner, M. G. Brik, P. Li, *Chem. Mater.* 27 (2015) 2938.

63. X. Huang, *Nat. Photonics* 10 (2014) 748.

64. J. Huo, W. Lü, B. Shao, Y. Feng, S. Zhao, H. You, *Dyes Pigments,* 139 (2017) 174.

65. S. V. Eliseeva, J. C. Bunzli, *Chem. Soc. Rev.* 39 (2010) 189.

66. Y. Cui, T. Song, J. Yu, Y. Yang, Z. Wang, G. Qian, *Adv. Funct. Mater.* 25 (2015) 4796.

67. C. Hang, J. Fei, Y. Tian, W. Zhang, C. Wang, S. Zhao, J. Caers, 14th International Conference on Electronic Packaging Technology Dalian, China (2013).

68. Moisture/reflow sensitivity classification for nonhermetic solid state surface mount devices, 2007 (IPC/JEDEC J-STD-020D).

69. Y. H. Lin, J. P. You, Y.-C. Lin, N. T. Tran, F. G. Shi, *IEEE Trans. Components Packag. Technol.* 33 (2010) 761.

70. Z. Ma, X. Zheng, W. Liu, X. Lin, W. Deng, Proc. 6th Int. Conf. Electron. Packag. Technol. Shenzhen, China, Sep. (2005) 1.

71. S. Nizamoglu, Fluorescent proteins for color-conversion light-emitting diodes, *SDÜ Fen Bilim. Enstitüsü Derg.* 20 (2016) 490.

72. M. D. Weber, L. Niklaus, M. Pröschel, P. B. Coto, U. Sonnewald, R. D. Costa, *Adv. Mater.* 27 (2015) 5493.

73. J. W. Hastings, K. L. Krause, Luciferases and light-emitting accessory proteins: Structural biology, in *Encyclopedia of Life Sciences.* Chichester, UK: Wiley (2006).

74. Y. H. Su, S.-L. Tu, S.-W. Tseng, Y.-C. Chang, S.-H. Chang, W.-M. Zhang, *Nanoscale* 2 (2010) 2639.

5

Phosphors for Photovoltaic Technology

5.1 Role of Rare Earth Phosphors in Photovoltaic Solar Cells

Today solar energy has become one of the most important sources of energy in the 21st century due to its universal existence, infinite reserve, and advantages of being clean and economic for sustainable development [1, 2]. A well known material that solar cells are made of is silicon and until recently there has been no substitute for Si-base solar cells. But, the photoelectric conversion efficiency of a conventional crystalline silicon solar cell is limited to 29% because of the spectral mismatch between the solar spectrum and the energy gap of silicon (1.12 eV) [3]. One of the major energy-loss mechanisms that affect the energy-conversion efficiency of Si-solar cells is the thermalization of the charge carriers, that are attributed via the absorption of high-energy photons [4, 5]. Recently, many efforts have been taken by multiple research organizations working in the field of photovoltaics to find substitutes for silicon that have good efficiency, or to find a way to increase the efficiency of existing C-Si-solar cell. Thus, among the variety of approaches aimed at raising the efficiency of the conventional photovoltaic (PV) cells, there is a specific approach for improving the process of energy harvesting. This is done by matching the wavelength of the solar spectrum and absorption wavelength of the PV cell not through active layer modification, but with the help of appropriate modifications of the solar spectrum. Therefore, this idea is well grounded for both down- and up- energy conversion processes that utilize the maximum wavelength range of solar light in energy conversion. This is promising in the 3rd generation of photovoltaics in the development of solar science and technology. Hence, to increase the efficiency of solar cells an optimal way is by reducing energy loss by converting a high-energy photon into two low energy NIR emitting photons by the process of down conversion [6–8]. Considering the energy levels of all lanthanides, depicted in the so-called Dieke energy level diagram, the rare earth ions such as Yb^{3+}, Er^{3+}, Nd^{3+}, Tm^{3+}, Sm^{3+}, Ho^{3+} etc. emits the light in visible and NIR (near infrared) ranges under UV excitations, and these ions have a relatively simple electronic structure which is formed by two energy-levels, from where the emission originates in broad or sharp bands separated by energy difference of a few electron volts that depends on impurity ions, and it is located just above the band gap of Si ($9090\,cm^{-1}$) [9]. Thus, rare earth ion–doped host materials are a promising candidates for the quantum cutting process and the use of downconversion phosphor to increase performance of Si-solar cells. It was recently studied theoretically by Truple et al. [10]. Truple reported the optimum solar cell band gap as nearly close to 1.1 ev. Hence, using a Si-solar cell with a coating of ideal DC phosphor, a conversion efficiency of the cell could be achieved of up to 38.6%. This is comparatively higher than the limiting efficiency of existing Si-solar cells i.e., 30.9%, without the use of downconversion phosphor layer. Thus, from the practical point of view more research on luminescence downconversion materials with external quantum efficiency of up to 90% has been proposed by many research groups for shifting the short wavelength to longer wavelength side. Therefore, the application of this QC/ downconversion layer in PV include luminescent solar concentration and gives more advantage to overcoming the limitation in the front surface of Si-solar cell design [11–13]. The energy loss can be reduced by modifying the solar spectrum via the downconversion process [14–16]. However, to improve the efficiency of Si-solar cell, the spectrum modification is the main challenge for the researcher, thus it is important to concentrate the study on new rare earth ion–doped NIR quantum cutting (QC) materials, which could have the capability to improve the energy conversion efficiency. Such downconversion or QC phosphor materials are a combination of rare-earth (RE) ions such as Tb^{3+}, Tm^{3+}, Pr^{3+}, Er^{3+}, Nd^{3+}, and Ce^{3+}) and Yb^{3+} ions mostly used as an activator for such processes, and they work on the mechanism that is based on downconversion energy transfer from one RE^{3+} ion to two nearby Yb^{3+} ions [17]. The RE^{3+} ions co-doped with Yb^{3+} ions are the favoured choice because Yb^{3+} ions emits light in $1000\,nm$ range and have simple energy levels, and RE^{3+} ions with a suitable energy level are approximately twice as large as the energy gap of the solar cell [18]. So, RE ion pairs of RE^{3+}–Yb^{3+} (RE = Tb, Pr, Tm) co-doped downconversion phosphors that can modify solar spectra by a cooperative energy transfer (CET) process and enhance the efficiency of photovoltaic cells [19–24]. But, the visible emission of absorption centres ions remains strong and the NIR luminescence of Yb^{3+}, in such a case, is still inevitably weak. This phenomenon of cooperative energy transfer is caused by the following reasons. First, the excitation is fixed at the corresponding wavelength of RE ions, for the Tb^{3+} ions, λ_{ex} must be constant at 308, 360, and $488\,nm$ respectively. And the absorption transitions of Tb^{3+}, Pr^{3+} and Tm^{3+} ions are attributed to intra-4f forbidden transitions (low absorption cross sections) typically of the order of $10^{-21}\,cm^2$. Whereas the upconversion process occurs in these RE^{3+} ions (Tb^{3+}, Pr^{3+}, Tm^{3+}) when it co-doped with Yb^{3+} ions, under the effective excitation by the energy transfer from Yb^{3+} ion (excited by NIR 971 nm laser) [25, 26], hence it is called a *back energy transfer* and it is energy efficient. Therefore due to such a drawback RE ions of this kind will not be useful for downconversion purposes in silicon-based solar cells. Generally, Ce^{3+} ion might be an ideal broadband sensitizer used in combination with with Yb^{3+} ion [27–30] in many host lattices due to the following reasons,

1. The absorption transition of Ce^{3+} is an allowed electric dipole transition from 4f ground state to 5d excited state, covering a broad spectral band.
2. The probability of back energy transfer (ET) from Yb^{3+} to Ce^{3+} is small.
3. The 4f–5d transition of Ce^{3+} can be tuned by host materials.

Thus, by choosing a suitable host material, downconversion from Ce^{3+} to Yb^{3+} can be enhanced, and this will have the good result of increasing the efficiency of Si cells. Among the different rare earth–activated ions, Eu^{3+} behaves as an efficient red-emitting activator, and it is used in many host materials [32–33]. The energy of the $^5D_2-^7F_0$ transition of Eu^{3+} is approximately twice as large as that of the Yb^{3+} $^2F_{5/2}-^2F_{7/2}$ transition, and hence, ET may be possible between Eu^{3+} and Yb^{3+} ions [34]. To date, only a few reports have been published regarding the downconversion luminescence properties and ET mechanism for $Eu^{3+}-Yb^{3+}$ ions [35–37]. Also, $CaSc_2O_4$ is a promising host for efficient upconversion/downconversion processes [38] due to its low-energy phonons [39], the short distances between positions that can be occupied by the dopants, and the high solubility of ytterbium ions. Efficient visible upconversion luminescence was obtained when the $CaSc_2O_4$ matrix co-doped with Er^{3+}/Yb^{3+} [40], Tm^{3+}/Yb^{3+} [41], and Ho^{3+}/Yb^{3+} ions, respectively [42]. Further, the mechanisms of cooperative downconversion and near infrared (NIR) luminescence were observed in $CaSc_2O_4$ phosphor doped with Tm^{3+}/Yb^{3+} ions [43].

5.2 Types of Phosphor

Phosphors are solid luminescent materials doped with activator ions. The best-known examples are copper-activated zinc sulfide and silver-activated zinc sulfide. Different types of phosphor host have been available for the past few decades: typically oxides, nitrides, oxynitrides, sulfides, halides, silicates, aluminum, and silicon, as discussed in Chapter 2. The activators used for doping purposes prolong the emission time. With double doping of rare earth ions, we can study the afterglow behavior of materials for display applications. Phosphors can emit light either through incandescence, in which all atoms radiate light, or by luminescence, in which only a small fraction of the atoms create emission centers or luminescence centers and emit light. In inorganic phosphors, these inhomogeneities in the crystal structure are created by the addition of small amounts of dopant impurities called *activators.* So, the doping ions create dislocations or other crystal defects in the host lattice. The wavelength emitted by the emission center is dependent on the atom itself and also on the surrounding crystal structure. The luminescence efficiency of rare earth–activated oxide-based phosphors is relatively low due to the high multiphonon relaxation rate caused by the high-energy photons. But oxide-based phosphor materials are more appropriate for practical applications because of their high chemical durability and thermal stability [44, 45, 46]. When rare earth ions are doped into a crystal lattice, the conversion efficiencies of the phosphor materials depend on the rare earth ion concentration, ET processes between host and doping ions, and the multiphonon relaxation rate. Therefore, today, there is a need for a rigorous search for novel rare earth–doped hosts having low-energy photons, high quantum efficiencies, and optimum doping concentration to develop optical devices with excellent luminescence properties [47–49].

5.2.1 Upconversion Phosphors

The general operating mechanism of upconversion phosphors has already been discussed in Chapter 2. Upconversion is a process in which the sequential absorption of two or more photons leads to the emission of light at a shorter wavelength than the incident wavelength [50]. Nicolaas Bloembergen et al. first proposed this mechanism, suggesting the photon counting abilities of lanthanide ions. It has diverse applications in many areas of technology, areas, such as energy conversion and medicine. Two low-energy photons excite an ion into a higher energy level, and a single high-energy photon is emitted when the ion makes the transition to the ground state. Energy transfer upconversion (ETU) is a process in which ET occurs between two rare earth ions (e.g., Yb^{3+}, Er^{3+}) with the release of high-energy photons. Upconversion shows an anti-Stokes-type emission, which converts infrared (IR) light into visible light [51–55]. Materials that show upconversion properties include the rare earth elements from the d-block or the f-block. Examples of these ions are Ln^{3+}, Ti^{2+}, Ni^{2+}, Mo^{3+}, Re^{4+}, Os^{4+}, and so on. Figure 5.1 shows the upconversion phosphor mechanism when sunlight is incident on the solar cell.

5.2.1.1 Some Reported Upconversion Phosphors

5.2.1.1.1 CaSc₂O₄ Phosphor Doped with Er³⁺ and Yb³⁺ Ion

$CaSc_2O_4$ is a promising host for an efficient upconversion/downconversion process due to its low-energy phonons [56–58]. This phosphor shows an intense visible upconversion luminescence in $CaSc_2O_4$ phosphor co-doped with Er^{3+}, Yb^{3+} [59], Tm^{3+}, Yb^{3+} [56], and Ho^{3+} and Yb^{3+} ions [59, 60]. Also, cooperative downconversion and NIR luminescence is observed in $CaSc_2O_4$:Tm^{3+}, Yb^{3+} phosphor [61]. When $CaSc_2O_4$ is doped with Ce^{3+} [62], Tb^{3+} [63], and Eu^{3+}

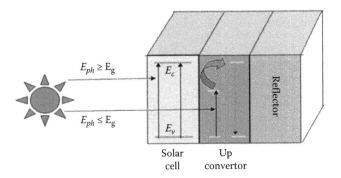

FIGURE 5.1
Upconversion phosphor mechanism.

[64], it shows efficient phosphor emission. Ştefan et al. successfully studied the radiative lifetimes of the main emitting levels of Ho^{3+}, Tm^{3+}, and Er^{3+} in a $CaSc_2O_4$ host lattice, which allowed them to calculate the quantum efficiencies of these levels. From the literature, it is seen that there is only a few published work on the upconversion luminescence in $CaSc_2O_4$ doped with Er^{3+}, Yb^{3+} ions [58], whereas Ştefan et al. [75] prepared $CaSc_2O_4$:Er^{3+}, Yb^{3+} nanopowders by the combustion method. Here, the doping concentrations of the rare earth were taken as $Yb^{3+}=6.0\,mol\%$ and $Er^{3+}=1.0\,mol\%$, respectively. Also, Ştefan successfully demonstrated the analysis of the upconversion properties of $CaSc_2O_4$:Er^{3+}, Yb^{3+} phosphors prepared by the solid-state method. $CaSc_2O_4$ has a calcium ferrite structure with space group P_{nam}, D^{16}_{2h} with three cationic site positions, in which two Sc^{3+} sites have six-fold coordination and one Ca^{2+} site has eight-fold coordination [65]. The position of the cations corresponds to C_s point symmetry. Also, the ionic radius of Sc^{3+} is $0.745\,Å$ and that of Ca^{2+} is $1.12\,Å$ [66]. Small trivalent rare earth ions such as Yb^{3+} ($0.868\,Å$) have six-fold coordination and preferentially occupy the Sc^{3+} sites [57, 67]. Also, Er^{3+} ions, having the ionic radius $0.89\,Å$ with six-fold coordination and $1.003\,Å$ with eight-fold coordination [66], can enter both Sc^{3+} and Ca^{2+} sites, respectively. In this case, Er^{3+} ions enter the Ca^{2+} cation sites, and hence, charge compensation is necessary for a calcium niobium gallium garnet (CNGG) or $CaWO_4$ host lattice [68, 69]. The charge compensation can be done only by a random distribution of impurities or by vacancies inside the host lattice. When $CaSc_2O_4$:Er^{3+}, Yb^{3+} phosphor is pumped at 980 nm, it shows a bright green wavelength range extending from 500 to 600 nm, corresponding to the $(^2H1_{1/2}, {}^4S_{3/2})–{}^4I_{15/2}$ transition, and red emission originating from the $^4F_{9/2}–^4I_{15/2}$ transition extending from 625 to 725 nm. The reported upconversion spectra of $CaSc_2O_4$:$Er_{(0.5\%)}$, $Yb_{(y)}$ (where $y=0\%–10\%$) phosphor for 124 mW incident pump power are depicted in Figure 5.2, and Figure 5.3 shows the UV-violet upconversion spectra for the $CaSc_2O_4$:$Er_{(0.5\%)}$, $Yb_{(5\%)}$ phosphor; here, the two upconversion-pumped luminescence transitions

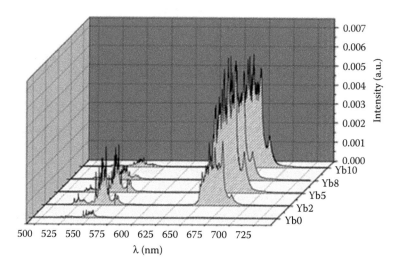

FIGURE 5.2

Visible upconversion spectra of $CaSc_2O_4$:Er (0.5%): Yb (y %) for $y = 0$, 2, 5, 8, 10. Incident pump power $P = 124$ mW. (With permission from Elsevier: *Journal of Luminescence* 180, (2016), 376, A. Stefan, O. Toma, S. Georgescu.) [75]

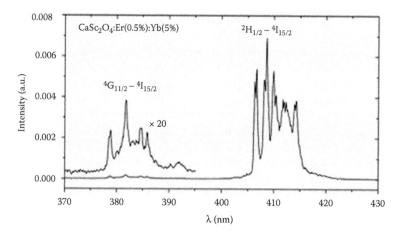

FIGURE 5.3

UV-violet upconversion spectrum for the $CaSc_2O_4$:Er(0.5%): Yb(5%) sample. (With permission from Elsevier: *Journal of Luminescence* 180, (2016), 376, A. Stefan, O. Toma, S. Georgescu.) [75]

are shown, and these are observed at ~380 nm, corresponding to $^4G_{11/2}$–$^4I_{15/2}$, and $^2H_{9/2}$–$^4I_{15/2}$, corresponding to 405 nm [75]. Also, the authors reported the UV-violet upconversion spectrum for $CaSc_2O_4$:Er$_{(0.5\%)}$, Yb$_{(5\%)}$ phosphor, as shown in Figure 5.3, for the same incident pump intensity; from this, it is seen that the observed luminescence bands are less intense than the green and red bands. When the pumping is applied at 980 nm, the energy levels corresponding to $^4F_{9/2}$, $^4S_{3/2}$, and $^2H_{11/2}$ are populated by two-photon processes,

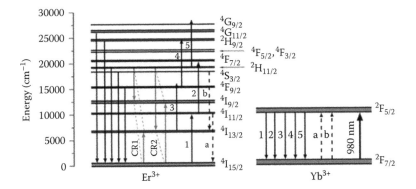

FIGURE 5.4
Energy level diagram of Er³⁺ and Yb³⁺ including their main energy transfer processes: energy transfer, back transfer, cross-relaxation processes. (With permission from Elsevier: *Journal of Luminescence* 180, (2016), 376, A. Stefan, O. Toma, S. Georgescu.) [75]

whereas the $^2H_{9/2}$ and $^4G_{11/2}$ levels are populated by three-photon processes. From the reported energy level schemes of Er³⁺ and Yb³⁺ lanthanide ions, as depicted in Figure 5.4. It is seen that the $^4G_{11/2}$ level is populated by the ET process. Also, the $^2H_{9/2}$ energy level is populated by the ET process due to ET from the $^4F_{9/2}$ level. The energy gap between the Er³⁺ levels of $^4G_{11/2}$ and $^2H_{9/2}$ is small—they are separated by an energy difference of 1400 cm⁻¹; thus, the multiphonon transition from $^4G_{11/2}$ may be responsible for contributing to the population of the 2/2 level [70, 71]. Ştefan et al. also reported that in the absence of Yb³⁺ ion, under 980 nm pumping, the excited-state absorption corresponds to the $^4I_{11/2}$–$^4F_{7/2}$ transition, and ETU is assigned to the $(^4I_{11/2}, ^4I_{11/2})$–$(^4F_{7/2}, ^4I_{15/2})$ transition; therefore, they contribute to the population of the $^4F_{7/2}$ level due to rapid multiphonon transition to populate the $(^2H_{11/2}, ^4S_{3/2})$ green-emitting levels. Here, the $^4F_{9/2}$ level is populated mainly due to the multiphonon transitions from $^4S_{3/2}$. The gray arrows in Figure 5.4 show the cross-relaxation (CR) processes involving the ground level $^4I_{15/2}$ assigned to CR1 and CR2, respectively, inside the Er³⁺ ion [72, 73]. Thus, the effective lifetime of the green-emitting level decreases. Therefore, with varying concentrations of Yb³⁺ co-doped ion, the intensities and the intensity ratios of green and red luminescence change as depicted in Figure 5.2. Similarly, if the Yb³⁺ concentration changes and the Er³⁺ concentration remains constant, the kinetics of the Er³⁺ levels becomes more rapid because of the back-transfer process, as shown in Figure 5.4.

5.2.1.1.2 Er³⁺/Yb³⁺ Co-Doped BaBi₂Nb₂O₉ Phosphor

Recently, Façanha et al. reported the PL properties of the Er³⁺/Yb³⁺ co-doped BaBi₂Nb₂O₉ phosphor prepared by the solid-state method [81]. The phosphor showed emission bands as represented in Figure 5.5. The phosphor was first excited at room temperature under the solid state laser excitation at λ = 980 nm,

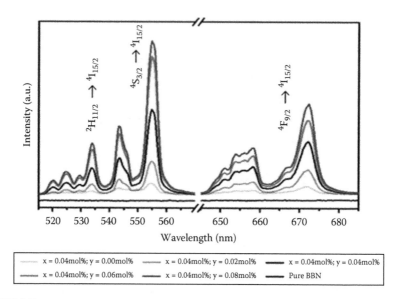

FIGURE 5.5
Upconversion emission spectra of system BBN: Er^{3+}, Yb^{3+}, excited at 980 nm. (With permission from Elsevier: *Journal of Luminescence*, 83, (2017), 102, M.X. Facanha, J.P.C. do Nascimento, M.A.S. Silva, M.C.C. Filho, A.N.L. Marques, A.G. Pinheiro, A.S.B. Sombra.) [81]

which power sets maintain at 0.429 and 13.6 mW/cm^2 respectively; under these conditions, the authors reported the emission spectrum for the pure BaBi$_2$Nb$_2$O$_9$ phosphor. But it does not show a PL process. The observed spectra reveal that the three distinct upconversion bands were located at 517 and 680 nm, of which the bands at 525 and 550 nm both correspond to the emission of green light assigned to the $^2H_{1/2}$–$^4I_{15/2}$ and $^4S_{3/2}$–$^4I_{15/2}$ transitions, respectively [75, 76–78], while the other peaks were reported at around 660 nm, corresponding to the red light emission and assigned to the $^4F_{9/2}$–$^4I_{15/2}$ transition [77–79]. The transitions discussed belong to the 4f–4f electronic transitions of Er^{3+} ions. Therefore, the emission spectrum of BaBi$_2$Nb$_2$O$_9$:$_{(0.04\,mol\,\%)}$ Er^{3+} $_y$Yb^{3+} phosphor clearly shows that the co-doped samples have higher emission intensities than the singly doped Er^{3+} ion. The composition BaBi$_2$Nb$_2$O$_9$:$_{(0.04\,mol\%)}$ Er^{3+} reports the most intense upconversion luminescence, when the concentration of impurity ion varies from $x = 0.005$ to 0.01, 0.02, and 0.04 mol%. Here, the upconversion emission intensity of samples varies as Yb^{3+} concentration increases. Also, the most intense emission peaks were observed at 525, 550, and 660 nm, as shown in Figure 5.5. The increase in the concentration of sensitizer ion promotes a continuous increase in the intensities of all upconversion bands, especially in the dominant-green emission (550 nm) between $y = 0.02$ mol% and $y = 0.06$ mol%, mostly due to interactions between doping ions. Therefore, it is concluded that the BaBi$_2$Nb$_2$O$_9$ sample doped with Er^{3+}/Yb^{3+} ions showed the most intense upconversion emissions. With further increase in the Yb^{3+} concentration, the PL intensity ratio of green/red band increases, although it is reported that the efficiency of the luminescence process generally decreases at higher dopant concentrations.

5.2.1.1.3 Upconversion Spectra of CeO_2:Er^{3+}, Li^+ Translucent Ceramics

Er^{3+} rare earth ion is well known for its emission characteristics from the NIR to the visible region, and it is mostly used in upconversion phosphors [81–85]. Wang et al. [97] reported that CeO_2:Er^{3+} gives bright visible upconversion emission, which can be clearly seen by the naked eye, having the laser excitation wavelength at 980 nm, as shown in Figure 5.6. The upconversion emission spectra of CeO_2:Er^{3+} and CeO_2:Er^{3+}, Li^+ translucent ceramics exhibit bands extending from 520 to 570 nm in the intense green region and red emission bands in the range of 640–680 nm, which are assigned to the $^2H_{11/2}$, $^4S_{3/2} \rightarrow {}^4I_{15/2}$, and $^4F_{9/2} \rightarrow {}^4I_{15/2}$ transitions of Er^{3+} ions, respectively. Also, the authors reported the effect of Li^+ doping on the emission intensity of CeO_2:Er^{3+} ceramics excited by a 980 nm diode laser, as depicted in Figure 5.6. During this investigation, they found that with increasing Li^+ ion doping concentration, the emission intensity also increased significantly. This enhancement in upconversion emission intensity may be attributed to the following aspects.

1. Substitution of Li^+ ions into the host would form a defect in the host lattice and create an oxygen vacancy to maintain charge neutrality; also, it acts as a sensitizer to transfer their energy to the doping ion [86].

2. Due to the smaller ionic radius of Li^+ ion (0.076 nm), which has a greater ability to substitute into the host crystal lattice of CeO_2, it also helps to enhance the local lattice field surrounding the Er^{3+} ion

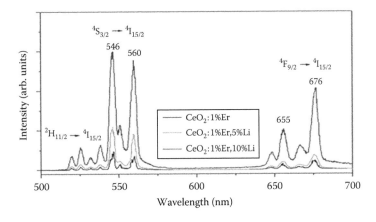

FIGURE 5.6
Upconversion emission spectrum of CeO_2:Er and CeO_2:Er^{3+}, Li^+ translucent ceramics under 980 nm diode laser excitation. (Reprinted from *Opt. Mater.*, 42, Guo, Y, et al., 390, Copyright (2015), with permission from Elsevier.) [97]

and lower its local symmetry; therefore, the lifetimes of the $^4I_{11/2}$ and $^4I_{13/2}$ energy levels of Er^{3+} ions are higher than that of a Li^+-free CeO_2:Er^{3+} sample, leading to an enhanced population of $^4I_{11/2}$ and $^4I_{13/2}$ levels of Er^{3+} ions, which improves the upconversion efficiency of the phosphor [87–90].

The formation of defects in the host lattice, such as pores, grain boundary, and grain defects, may be remarkably reduced by adding Li^+ ions during the sintering process, which also decreases the luminescence quenching centers in ceramics samples, and hence, the radiative transition rate of Er^{3+} ions increases. Thus, the Er^{3+} upconversion luminescence in ceramics samples is enhanced more strongly. To resolve more accurately the upconverting mechanism of CeO_2:Er^{3+} and CeO_2:Er^{3+}, Li^+ ceramics, the upconverting emission intensity I of Er^{3+} ions as a function of the pumped power P was monitored, and the intensity I was found to be proportional to the nth power of P, that is, I/P, where n is the ratio of the number of photons absorbed during pumping per upconverted emitted photon [91]. Therefore, the graph between log I versus log P yields a straight line having the slope n. Its value is 1.99 and 1.98 for the 520–570 nm and 640–680 nm upconverted emissions range of the CeO_2:1% Er, 5% Li ceramics sample, respectively. Therefore, the two-photon upconversion emission process is involved in this case. The energy level diagram and the upconversion mechanism of Er^{3+} ion were proposed and discussed well by Wang et al., as depicted in Figure 5.7, for the intense green and red emissions excited at 980 nm [92–96]. The Er^{3+} ions present in the ground state absorb the 980 nm excitation wavelength and go to the $^4I_{11/2}$ level, and

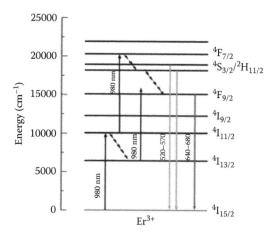

FIGURE 5.7

Upconversion mechanism in CeO_2:Er, Li translucent ceramics under 980 nm diode laser excitation. (Reprinted from *Opt. Mater.*, 42, Guo, Y., et al., 390, Copyright (2015), with permission from Elsevier.) [97]

this state relaxes to the $^4I_{13/2}$ level through a multiphonon nonradiative relaxation process, thus increasing the population of the $^4I_{13/2}$ level. Again, the Er^{3+} ion residing on the $^4I_{11/2}$ level is further excited to the $^4F_{7/2}$ level through the excited-state absorption process. Therefore, this nonradiative relaxation could populate the $^2H_{11/2}$ and $^4S_{3/2}$ levels, from which the green upconversion luminescence originates.

5.2.1.1.4 Ho³⁺-Doped BaYbF₅ Nanocrystals

Recently, many researchers have focused on rare earth–activated transparent oxyfluoride glass-ceramics (GCs) as interesting luminescent materials because of their excellent luminescence properties [98–109] from the IR to the visible region of the spectrum. These fluoride hosts emit a low-cut-off phonon with desirable mechanical and chemical characteristics. Among the different lanthanides, Ho^{3+} is well known for its observed optimal characteristic emission bands, because of its abundant energy levels ranging from NIR to ultraviolet, wherein a few of the energy levels are separated by a gap of $1000\,cm^{-1}$ [110], and it is used as a potential candidate for a solid-state upconversion laser [111–113]. The upconverted luminescence in these ions can only be observed in fluoride chalcogenide and halide crystal [114]. Thus, $BaYbF_5$ crystal is a promising host material for upconversion [115]. Guo et al. reported a detailed analysis of fabrication, structural characterization and upconversion properties of novel Ho^{3+}–doped $BaYbF_5$ nanocrystals [116]. Figure 5.8a represents a PL spectrum of $BaYbF_5{:}Ho^{3+}$ nanocrystals extending from 500 to 800 nm; the different bands were assigned to the transition from $^5F_4/^5S_2-^5I_8$, which exhibits green emission; $^5F_5-^5I_8$ belongs to red emission and $^5F_4/^5S_2-^5I_7$ corresponds to NIR emission. The inset in Figure 5.8 shows that there are three different types of upconversion spectra for GC660, GC640, and precursor glass (PG); from this, it is reported that upconverted emissions are greatly enhanced in the case of the GCs. The GC660 sample shows stronger upconversion emission, with clear Stark splitting, than the other samples. Thus, from these comparative PL spectra, it is seen that with PG, the green to red emission intensity ratio increases by about 20, whereas it increases 40 times for GC640 and GC660. This results in the color varying from reddish (PG) to green (GC660). The upconverted emissions spectrum of Ho^{3+} is reported at 362, 388, 418, 445–456, and 485 nm as shown in Figure 5.8b for GC, but it is absent in the case of the PG sample. Therefore, the observed green (538 nm) and red (650 nm) emissions correspond to two-photon processes, while the blue emission (485 nm) is due to the three-photon process. By considering the energy level diagram of Ho^{3+} and Yb^{3+} ions in crystal, it is concluded that in this case, the photon can be pumped by 980 nm, which can be efficiently absorbed by Yb^{3+} ions, and hence, Ho^{3+} levels ($^5F_5$ and $^5S_2/^5F_4$) are populated through two successive ETs from $Yb^{3+} \longrightarrow Ho^{3+}$ ions, respectively.

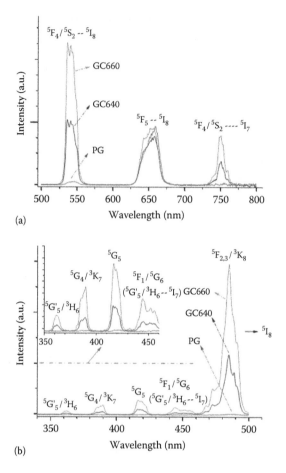

FIGURE 5.8
Upconversion spectra of PG, GC640, and GC660:Ho^{3+} samples (λ_{ex}=980 nm), (a) Photo-luminescence spectra of BaYbF$_5$: Ho^{3+} nanocrystals, (b) Upconverted emissions spectra of Ho^{3+} ion. (Reprinted from *J. Lumin.*, 152, Jiang, S., et al., 195, Copyright (2014), with permission from Elsevier.) [116]

5.2.2 Downconversion Phosphor

A preliminary idea of the working of downconversion phosphors has been discussed in Section 3.5.2 of Chapter 2. In recent years, rare earth–doped inorganic luminescent materials have been used as a downconversion layer for solar cells, because they have the capability of shifting or converting a broad spectrum of light into photons of longer wavelength [117, 118]. Thus, downconversion is a process involving more than one photon, in which one absorbed high-energy photon is converted into more than one lower-energy photon. This process follows the Stokes law, with the change in wavelength known as the Stokes shift. Here, the energy loss is minimized by the thermalization of hot charge carriers after the absorption of high–incident energy photons [119, 120]. This downconversion layer can be placed on the front side of a solar cell as a light absorption layer, which has the ability to generate light of a more suitable wavelength to be absorbed by the photo-sensitive layer. It is widely used in silicon solar cells, dye-sensitized solar cells, and quantum dot–sensitized solar cells to improve the light conversion efficiency of the cell [121–129].

5.2.2.1 Some Reported Work

5.2.2.1.1 SrAl₂O₄:Eu²⁺/Yb³⁺ Downconversion Phosphor

Recently, Tai et al. reported the PL excitation and emission spectra of a $SrAl_2O_4:Eu^{2+}$ phosphor observed over a broad spectral range, in which the emission band was examined at 515 nm, under 370 nm excitation wavelength. Here, the broad bands were assigned to the 5d–4f transition of Eu^{2+} ion [130]. Thus, isolated Eu^{2+} bands were obtained in the present sample. Further, the author compares the PL spectra of $SrAl_2O_4$: Eu^{2+}, Yb^{3+} phosphor as shown in Figure 5.9a, b. Therefore, the mechanism of ET is followed from Eu^{2+} to Yb^{3+} ions effectively due to the resonant ET condition of donor and accep-tor ions. The PL spectra of $SrAl_2O_4:Eu^{2+}$, Yb^3 phosphor would be reported in the broad excitation range by keeping Yb3+ ion emission constant at 980 nm and similarly the excitation spectrum of Eu2+ obtained by monitoring the emission wavelength constant at 515 nm. Further, the charge transfer state (CTS) excitation bands of both the Eu^{2+} and $Yb^{3+}–O^{2-}$ peaks are at the same

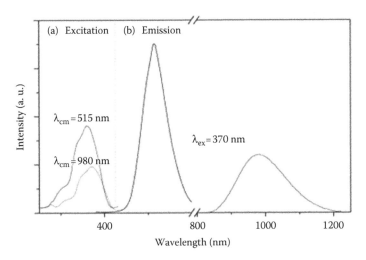

FIGURE 5.9
(a and b) PL spectra of $SrAl_2O_4:Eu^{2+}$, Yb^{3+}, where (a) shows a PLE spectrum of $SrAl_2O_4{:}_{0.01}Eu^{2+}$, $_{0.01}Yb^{3+}$ (λ_{em}=515 and 1064 nm), (b) PL emission spectrum of $SrAl_2O_4:Eu^{2+}$, Yb^{3+} (λ_{em} = 370 nm). (With permission from Elsevier: *Journal of Solid State Chemistry*, 226, (2015), 250, Y. Tai, G. Zheng, H. Wang, and J. Bai.) [130]

position, that is, 264 nm, when the NIR Yb^{3+} emission is kept constant, which is attributed to electron transfer from the $2p^6$ orbital of O^{2-} to the $4f^{13}$ orbital of Yb^+ ion [131]. Also, the authors presented ET from Eu^{2+} to Yb^{3+} on the basis of PL emission spectra; from Figure 5.9b, it is seen that $SrAl_2O_4$ phosphor co-doped with Eu^{2+} and Yb^{3+} shows an emission band at 980 nm for Yb^{3+} and at 515 nm for Eu^{2+}, keeping the excitation wavelength constant at 370 nm. The observed excitation spectrum of $SrAl_2O_{4:0.01}Eu^{2+}$, Yb^{3+} was further resolved by using the curve fitting analysis, in which two bands were clearly observed: one was assigned to CTS of the Yb–O bond, and another strong band corresponded to the Eu^{2+} emission, as depicted in Figure 5.10. Also, it is noticed that the presence of CTS between Yb^{3+} and O^{2-} did not affect the ET process from Eu^{2+} to Yb^{3+} ion under the 370 nm excitation wavelength, as discussed by Lin et al. [132]. This observed CTS at 264 nm corresponds to the energy level of 38,000 cm^{-1}, which was located at a higher energy level than the Eu^{2+} band, which had an energy level of 27,000 cm^{-1}. Thus, the CTS of Eu^{2+} required the assistance of phonons, and the difficulty occurred due to the large energy gap between Eu^{2+} 5d excited state and CTS occurring at 11,000 cm^{-1}. Therefore, these results confirmed that Eu^{2+} emission bands were dominated by the ET process. Also, it was mentioned that no Eu^{3+} emission was found in the observed results under 370 nm excitation wavelength, and the CTS between Yb^{3+} and O^{2-} was negligible. Figure 5.11 shows the reported emission spectra of $SrAl_2O_{4:0.01}$ Eu^{2+}, Yb^{3+} phosphor with varying Yb^{3+} content in the visible and NIR regions of S_2, S_3, S_4, S_5, and S_6 excited at 370 nm wavelength. Thus, in the case of S_2 to S_5, the Yb^{3+} concentration increased from 2 to 20 mol%, whereas the Eu^{2+} emission intensity at 515 nm monotonically decreased when the Yb^{3+} concentration rapidly increased. This gives

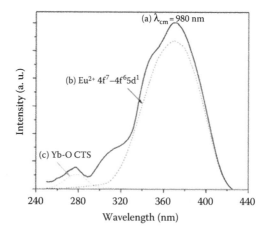

FIGURE 5.10
(a) PLE spectra of $SrAl_2O_4:Eu^{2+}$, Yb^{3+} phosphor at 980 nm NIR emission, (b) Eu^{2+} emission by Gaussian fitting, (c) Yb–O CTS by Gaussian fitting. (With permission from Elsevier: *Journal of Solid State Chemistry*, 226, (2015), 250, Y. Tai, G. Zheng, H. Wang, and J. Bai.) [130]

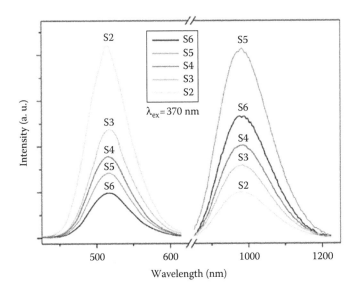

FIGURE 5.11
PL emission spectra of SrAl$_2$O$_4$: Eu^{2+}, Yb^{3+} with varying Yb^{3+} content on excitation of Eu^{2+} at 370 nm. (With permission from Elsevier: *Journal of Solid State Chemistry*, 226, (2015), 250, Y. Tai, G. Zheng, H. Wang, J. Bai.) [130]

strong evidence that the ET from Eu^{2+} to Yb^{3+} ion was dominated by the Eu^{2+} transition only. From this observation, the authors reported the PL properties of S$_0$ and S$_1$ and successfully showed NIR quantum cutting (QC) by the CET process from Eu^{2+} to Yb^{3+} ion. Thus, the observed results revealed efficient conversion of UV to the visible broadband range (250–450 nm), which is not used by solar cells, so that it is first converted to NIR (980 nm light, which can be fully absorbed by crystalline silicon).

5.2.2.1.2 Yb^{3+}-Doped CdWO$_4$ Downconversion Phosphor

Figure 5.12 shows the excitation and emission spectra of CdWO$_4$:Yb^{3+} (2%) nanophosphor, reported by Vanetsev et al., [133]. It is seen that the excitation bands' wavelength range of both Yb^{3+} and CdWO$_4$ host almost coincided with each other, but the emission spectra for host and dopant are not the same. Thus, the excitation bands were centered at 290 nm for pure CdWO$_4$, and the emission corresponded to 475 nm, whereas after doping with Yb^{3+} ion, the emission bands were reported in the IR region, at 980 nm, under the same excitation wavelength. Here, Yb^{3+} ion was effectively excited at 350 nm, which indicates the conversion of absorbed UV photons into Yb^{3+} IR fluorescence due to the ET from the CdWO$_4$ phosphor matrix to Yb^{3+} ion. Thus, from the reported results, it is seen that there is a conversion of UV incident photons, which are absorbed near the intrinsic absorption edge of CdWO$_4$, to IR emission of doping Yb^{3+} ions. Therefore, during this conversion mechanism, the following process must be involved.

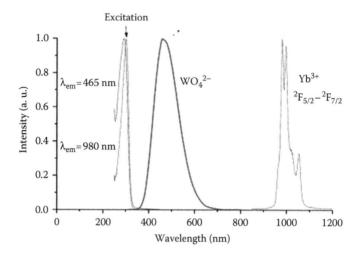

FIGURE 5.12

PLE and emission spectra of CdWO$_4$:Yb^{3+} (2%) nanophosphor, under λ_{exc} = 290 nm. (Reprinted from *Radiat. Meas.*, 90, Vanetsev, A. S. et al., 329, Copyright (2016), with permission from Elsevier.) [133]

1. So, the first mechanism is the recombination process, in which the Yb^{3+} ion captures an electron and a hole when absorbed by the host.

2. There is an ET from a nonrelaxed exciton to Yb^{3+}–O$_2$ CTS with subsequent nonradiative relaxation to the $^2F_{5/2}$ excited level of Yb^{3+} ion.

3. Finally, Yb^{3+}–O$_2$ CTS is created in the spectral region near the absorption edge of CdWO$_4$.

Thus, in this work, the author concluded that luminescence studies from the synthesized CdWO$_4$:Yb^{3+} nanopowders have shown a possible mechanism of downconversion in which the absorbed UV photons are converted into IR fluorescence of Yb^{3+} ions. Figure 5.13 shows the ET process from WO$_4^{2-}$ to Yb^{3+} ions, which shows that incident UV photons are first absorbed by WO$_4^{2-}$ and shifted to an excited level. Then, through a nonradiative process, they return to the metastable state with the emission of blue-green light, and further, they transfer their energy to the upper excited level of Yb^{3+} ions with emission of IR light.

5.2.2.1.3 *ZrO$_2$-Doped Eu^{3+}–Yb^{3+}–Y^{3+} Downconversion Phosphor*

Recently, Liao et al. [134] reported the excitation spectra of a Eu3+ -Yb3+ -Y3+ tri-doped cubic ZrO2 (abbreviated as YSZ:Eu,Yb), Yb phosphor keeping emission constant at 613 and 969 nm, as shown in Figure 5.14. Under constant emission at 613 nm, the YSZ:Eu^{3+} phosphor shows an absorption band at 285 nm, which is assigned to the charge-transfer (CT) transitions from O^{2-} to Eu^{3+} ion [135]. Also, a sharp 4f–4f transition of Eu^{3+} is observed

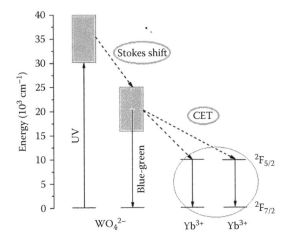

FIGURE 5.13
ET process from WO_4^{2-} to two Yb^{3+}. (Reprinted from *Radiat. Meas.*, 90, Vanetsev, A. S. et al., 329, Copyright (2016), with permission from Elsevier.) [133]

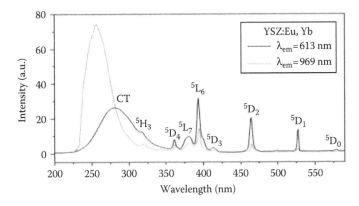

FIGURE 5.14
PLE spectra of Eu^{3+} emission (monitored at 613 nm) and Yb^{3+} emission (monitored at 969 nm) of YSZ:Eu^{3+}, Yb^{3+} samples.(From *Physica B*, 436, Liao, J., et al., 59, 2014, Copyright (2014) permission from Elsevier.) [134]

due to the transition from the ground state $^7F_0$ to the upper excited states $^5D_{0, 1, 2, 3, 4}$, $^5L_{6, 7}$, and $^5H_3$, as depicted in Figure 5.14. In this case, it is mentioned that the excitation spectra were clearly observed under the 969 nm emission wavelength of Yb^{3+} ion. The strong absorption band assigned to the CT transitions both from O^{2-} to Yb^{3+} and from O^{2-} to Eu^{3+}, which was located at 257 nm [135], helped to enhance the NIR intensity of Yb^{3+} ion. Thus, the ET from Eu^{3+} to Yb^{3+} ions in YSZ:Eu^{3+}, Yb^{3+} phosphor was possible only when there existed a strong CT absorption band with the presence of relatively weak absorption peaks of Eu^{3+} in the excitation spectrum of the Yb^{3+} ions.

The downconversion mechanism from Eu^{3+} to Yb^{3+} in a YSZ phosphor is successfully reported by the author, and it is shown in Figure 5.15. From the energy level diagram, it is seen that in the cooperative ET process, the energy of the $^7F_0-^5D_2$ transition of Eu^{3+} ion is approximately twice that of the $^2F_{5/2}-^2F_{7/2}$ transition of Yb^{3+} ion, and it does not correspond to other levels in the UV region. During the ET process, the $^5D_2$ excited state of the Eu^{3+} ion can transfer its energy to two nearby Yb^{3+} ions, after which the Yb^{3+} ions emit two IR photons. These processes are followed by rapid multiphonon-assisted relaxation from the populated levels to the $^5D_2$ level and finally, falling to the metastable state $^5D_0$, as depicted in Figure 5.15. ET from Eu^{3+} to Yb^{3+} might be involved in CR between the Eu^{3+}: $^5D_0-^7F_6$ (~12,000 cm^{-1}) and Yb^{3+}: $^2F_{7/2}-^2F_{5/2}$ (~10,000 cm^{-1}) transitions with the assistance of phonons. Hence, the photon in the $^5D_0$ excited state of the Eu^{3+} ion may transfer its energy to one Yb^{3+} ion, which assists approximately five phonons having the vibration energy difference of the host ions equal to 470 cm^{-1}; thus, in this case, only one IR photon is generated. In Figure 5.16, the author reported the emission spectra of YSZ:Eu^{3+}, Yb^{3+} corresponding to the visible region, from 550 to 720 nm, and the NIR region, from 900 to 1150 nm. Under 393 nm excitation wavelength, the emission peaks were located at 592 nm ($^5D_0-^7F_1$), 613 nm($^5D_0-^7F_2$), 651 nm ($^5D_0-^7F_3$), and 704 nm ($^5D_0-^7F_4$), and these bands originated from the f–f transition lines within the $4f^6$ electron configuration of Eu^{3+} ion. The author also depicted the integrated emission intensities of Eu^{3+} for the $^5D_0$ state, which shows that intensities of Eu^{3+} rapidly decrease as the Yb^{3+} concentration increases, as shown in the inset of Figure 5.16. As we go

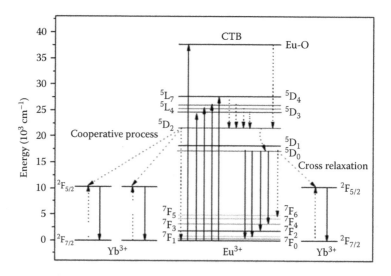

FIGURE 5.15
Schematic energy level diagram of Eu^{3+}/Yb^{3+} co-doped YSZ, showing the energy transfer process of NIR QC from the excited state of Eu^{3+} to the $^2F_{5/2}$ levels of Yb^{3+}. (From *Physica B*, 436, Liao, J., et al., 59, 2014, Copyright (2014) permission from Elsevier.) [134]

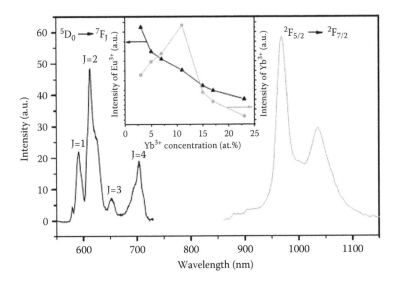

FIGURE 5.16
Visible and NIR emission spectra of Eu^{3+} and Yb^{3+} in a YSZ:Eu^{3+}, Yb^{3+} phosphor under the 393 nm excitation wavelength. (Inset shows the concentration of Yb^{3+} in $Eu_{0.06}$ $Y_{0.03}YbxZr_{(3.73-3x)}$ /$4O_2$ host, where $x = 3$, 7, 11, 15, 19, and 23 at %). From *Physica B*, 436, Liao, J., et al., 59, 2014, Copyright (2014) permission from Elsevier.) [134]

through the energy level diagram of Eu^{3+}/Yb^{3+} co-doped YSZ phosphor, it is found that two possible ET processes exist: one is CET, and the other is a CR from Eu^{3+} to Yb^{3+} ion [135]. The emission bands observed at 969 and 1035 nm were due to the transition of Yb^{3+} from the high $^2F_{7/2}$ multiplet to the low $^2F_{5/2}$ multiplet Stark level. Therefore, with varying concentration of Yb^{3+} ion, the observed NIR emission is also enhanced. This gives strong evidence of efficient energy transfer from Eu^{3+} to Yb^{3+} in the present host lattice, and thus, it may be used as a downconversion phosphor layer to enhance the efficiency of existing Si solar cells.

5.2.2.1.4 *LiGdF₄:Eu³⁺ Downconversion Phosphor*

To explain the mechanism of QC in a LiGdF4:Eu^{3+} host lattice, the relevant ET processes from $Gd^{3+} \longrightarrow Eu^{3+}$ ion were schematically represented by Meijerink et al. [136]. In the past, Feldmann et al. carried out a PL study of $LiGdF_4$ materials; Figure 5.17 presents the excitation spectra of the $LiGdF_4$:Eu^{3+} sample reported by Feldmann et al., in which the bands exhibit a series of sharp lines [137, 138]. The observed excitation bands were monitored by keeping the Eu^{3+} emission constant at 556 and 590 nm, and these bands were assigned to $^5D_1 \longrightarrow ^7F_2$ and $^5D_0 \longrightarrow ^7F_1$ transition, respectively; the authors also reported the presence of a QC mechanism in $LiGdF_4$:Eu^{3+} phosphor, which was detected by comparing the line intensities of the Eu^{3+} excitation. The strong absorption bands of Gd^{3+} observed at 273 nm correspond to the $^8S_{7/2} \longrightarrow ^6I_J$ transition. Again, it is mentioned that the intensity of Gd^{3+} excitation, which

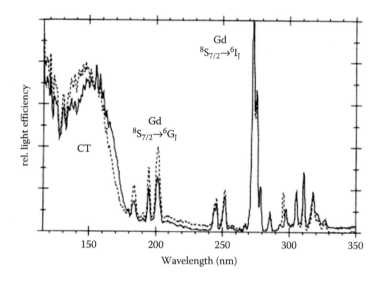

FIGURE 5.17

Excitation spectra of LiGdF4 sample keeping emission constant at 554 nm (solid) and emission at 590 nm (dashed). (From *J. Lumin.*, 92, Feldmann, C., et al., 245, 2001, Copyright (2001) permission from Elsevier.) [137]

appeared at 183, 195, and 202 nm, corresponded to the $^8S_{7/2} \longrightarrow {}^6G_J$ transitions with respect to those bands appearing at 554 and 590 nm, respectively. The presence of QC in this phosphor was clearly discussed on the basis of its PL emission spectra as reported by the author in Figure 5.18; it was compared by monitoring the excitation at 160 nm, corresponding to the ^6G_J levels of Gd^{3+} at 202 nm, and another excitation band at 273 nm, due to the ^6I_J levels of Gd^{3+}. The band appearing at $^5D_1 \longrightarrow {}^7F_2$ transition should be unaffected by the QC effect. The emission spectra are scaled with respect to the $^5D_1 \longrightarrow {}^7F_2$ transition, which would be unaffected by the QC effect, and due to this effect on excitation in the ^6G_J levels of Gd^{3+}, it is noted that the amplitude of the $^5D_0 \longrightarrow {}^7F_J$ transition was increased two-fold. Thus, based on the efficiency of the CR step, further, a 95% probability of CR is reached; therefore, the down-conversion efficiency was found to be 195%. In this work, the authors also reported that exciting the phosphor at 183, 195, or 202 nm, belonging to the $^8S_{7/2} \longrightarrow {}^6G_J$ transition of Gd^{3+}, does not give a two-photon emission; although another broad band around 160 nm also had no QC emission, indicating that nonradiative decay via lattice defects or ET to higher energy levels of Eu^{3+} produced a single-photon emission, which was more efficient. Figure 5.19 shows the PL emission spectra measured under the excitation at 202 nm. The spectral emission intensity was found to be 82% at 202 nm [137]. Based on an investigation under *vacuum UV* (VUV) excitation, the author stated that the $LiGdF_4:Eu^{3+}$ phosphor shows the QC mechanism only at 202, 195, and 185 nm excitation. Hence, it is concluded that research into the development of QC phosphors based on fluoride lattices should help in the improvement of the

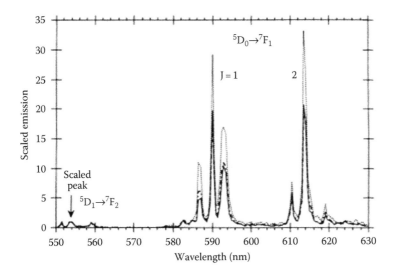

FIGURE 5.18
PL emission spectra of LGF sample on excitation at 160 nm (dashed line), on excitation in the ^6G_J levels of Gd^{3+} at 202 nm (dotted line), and on excitation in the ^6I_J levels of Gd^{3+} at 273 nm (solid line).(From *J. Lumin.*, 92, Feldmann, C., et al., 245, 2001, Copyright (2001) permission from Elsevier.) [137]

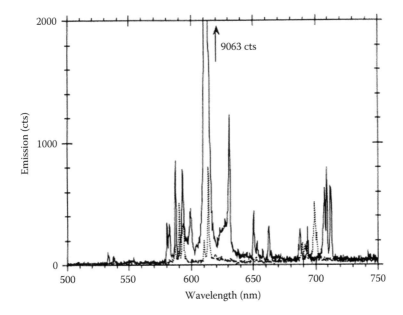

FIGURE 5.19
PL emission spectra of YOX (solid) and LGF (dotted) recorded under excitation at 202 nm. (From *J. Lumin.*, 92, Feldmann, C., et al., 245, 2001, Copyright (2001) permission from Elsevier.) [137]

crystallinity of the fluoride lattice, which has better UV/VUV reflectance and good quantum efficiencies.

5.3 Recent Developments in PV Technology

The conversion of solar energy into electricity was discovered by French physicist Alexander Edmond Becquerel in 1839. This was the beginning of solar cell technology. Solar cells can be broadly categorized into three generations. The first generation includes solar cells that are relatively higher in cost to fabricate and have low conversion efficiency. The second generation involves types of solar cells that have lower efficiency but are much cheaper to fabricate, and the solar cell cost per watt is lower than in first-generation cells. The third generation of solar cells are very efficient. Most technologies in this generation are not yet commercial; many research organizations are working on the development of highly efficient cells and the reduction of their production costs.

5.3.1 First Generation of Photovoltaics

Generally, first-generation solar cells include

1. Single-crystal solar cells
2. Multicrystal solar cells

Thus, first-generation cells are classified on the basis of their crystallization level. If the whole wafer is formed of only one crystal, it is referred to as a *single-crystal* solar cell. If the wafer is formed of crystalline grains, it is called a *multicrystal* solar cell. We can clearly see the boundaries between grains that exist in the solar cell. These are the oldest and the most commonly used technology due to their high efficiencies. They are produced on wafers in which each wafer can supply 2–3 watts of power. The power of the cell increases with the use of many cells in solar modules. We know that the efficiency of monocrystal cells is always higher than that of multicrystal solar cells. The fabrication and processing of multicrystal wafers are easier and cheaper. Currently, research in the field of wafer-based solar cells is concentrating on the development of new and advanced process technologies, such as texturization via nanoimprint lithography or the production of selective emitters by screen printing, reactive ion etching, laser doping, and so on. The development of this technology is accompanied by extensive device and simulation processing. Also, advanced optical and electrical measurement is being developed to study the characteristics of the solar cells.

Therefore, today's great challenge in PV technology is to

1. Increase the efficiencies
2. Reduce the production costs

Generally, first-generation solar cells are based on silicon (Si); to date, this technology has high conversion efficiency, but the fabrication cost of the cell is too high. Also, the process is complex, and several issues affect the efficiency of cells, including:

- The energy of the incident photons on the solar cell is lower than the band gap, so that most of the incident light cannot be used by the cell (because of spectral mismatch).
- The energy of the incoming photons is greater than the band gap; therefore, the excess energy is lost as heat.
- For n-type and p-type silicon, Fermi levels are always inside the band gap of silicon. Therefore, the open-circuit voltage is lower than the band gap.

5.3.2 Second Generation of Photovoltaics

The second generation contains a type of solar cells that have low efficiency but are much cheaper to produce, such that the fabrication cost per watt of the second-generation cell is lower than for the first generation. This class of cells is based on amorphous silicon, copper indium gallium selenide (CIGS), and cadmium telluride/cadmium sulfide (CdTe), and the cell performance is 10%–15%. It does not contain a silicon wafer, and it requires lower material consumption; therefore, it helps to reduce the production costs of this type of solar cells compared with the first generation. For this type, we can fabricate cells so that they are flexible to some degree. The production of second-generation solar cells still requires vacuum processes and high-temperature treatments; hence, there is high energy consumption during the production of such solar cells. Second-generation solar cells are based on scarce elements, and this is a limiting factor in the price. This type of cell includes amorphous thin film solar cells, CdTe/CdS solar cells, CIGS solar cells, and so on. Following are the advantages and disadvantages of second-generation thin film solar cells.

Advantages:

- High absorption coefficient.
- Can occupy both vacuum and non-vacuum processes.
- Lower cost in comparison with C-Si-based solar cells.
- Low-cost substrate (Cu tape).

Disadvantages:

- Environmental contamination starts with the fabrication process.
- Materials are hard to find.

5.3.3 Third Generation of Photovoltaics

The third generation of PV solar cells are potentially able to overcome the Shockley–Queisser limit of 31%–41% power efficiency for single–band gap solar cells. Commonly used third-generation systems include multi-layer ("tandem") cells made of amorphous silicon or gallium arsenide, while more theoretical developments include frequency conversion hot-carrier effects and other multiple-carrier ejection techniques, and so on.

Third-generation PVs are well known for their efficiency. Most technologies in this generation are not yet commercialized, but there is a need of a lot of research in this area to develop more efficient solar cells with low material costs that are relatively cheap.

Third-generation solar cells include

1. Nanocrystal-based solar cells
2. Polymer-based solar cells
3. Dye-sensitized solar cells
4. Concentrated solar cells

Since the third-generation approach is helpful to increase the efficiency of the new cell as compared with existing cells, research is underway on engineering wider band gaps for Si-based materials using quantum confinement in nanostructures. This can be achieved using either quantum wells (QWs) or quantum dots (QDs) of Si sandwiched between layers of a dielectric based on Si compounds such as SiO_2, Si_3N_4, or SiC_{21}. Here, for a sufficiently close spacing of QWs or QDs, a true miniband is formed, creating an effectively larger band gap.

For QDs of 2 nm or QWs of 1 nm, having an effective band gap of 1.7 eV indicates that they are ideal for a tandem cell element on top of Si. A thin layer is grown by thin-film sputtering or chemical vapor deposition processes followed by a high-temperature anneal to crystallize the Si QWs/QDs. The matrix remains amorphous, thus avoiding some of the problems of lattice mismatch; thus, it suffers from the problem of spectral sensitivity. Today, many researchers are working on silicon third-generation cells using the concept of nanotechnology, especially band gap engineering via quantum confinement effects, and this is being explored to improve Si-based cells. Using thin Si layers, vertical quantization yields a substantially increased band gap suitable for absorption of the short-wavelength portion of the solar spectrum without relaxation losses. A major challenge here is how to create ultra-thin Si layers with a high degree of crystallinity. Currently, research organizations are aiming to improve the crystalline quality of the multi-layer stacks, which would not only strongly increase the efficiency of third-generation cells but be interesting for photonics applications; these techniques are promising but not commercially proved. As mentioned earlier, the most developed third-generation solar cell types are dye-sensitized and concentrated solar cells. Dye-sensitized solar cells (DSSCs) are based on dye molecules between electrodes; the electron–hole pairs occur in dye molecules and are transported through TiO_2 nanoparticles. The efficiency of the cell is low, but the fabrication process is cheaper than Si-solar cells. Another class is the concentrated PV (CPV) solar cell, and this is a promising technology. The working principle of concentrated cells is to concentrate a large amount of solar radiation in a small region, where the PV cell is located. In this way, the amount of expensive semiconductor materials is reduced. CPVs are a promising PV technology in the near future.

Advantages of third-generation polymer-based solar cells:

- The raw materials are easy to find.
- The fabrication process is easier than for the other two technologies.
- Cost is minimal.

Disadvantages of third-generation organic-based solar cells:

Liquid electrolyte (low temperature)High cost of Ru (dye) and Pt (electrode)

References

1. PIKE Technologies Diffuse Reflectance—Theory and Applications Note (www.piketech.com/files/pdfs/PIKE_Diffuse-Reflectance-Theory-Applications.pdf.).
2. B. van der Zwaan, A. Rabl, *Sol. Energy* 74 (2003) 19.
3. O. Morton, *Nature* 443 (2006) 19.
4. W. Shockley, H. J. Queisser, *J. Appl. Phys.* 32 (1961) 510.
5. T. Trupke, M. A. Green, P. Würfel, *J. Appl. Phys.* 92 (2002) 1668.
6. B. S. Richards, *Solar Energy Mater. Sol. Cells* 90 (2006) 1189.
7. T. Trupkea, B. Mitchellb, J.W. Webera, W. McMillana, R.A. Bardosa, and R. Kroezea, Energy Procedia 15 (2012) 135.
8. B.S Richards, A Shalav, IEEE Transactions on Electron Devices 54 (2007) 2679.
9. Q.Y. Zhang, X.Y. Huang, *Prog. Mater. Sci.* 55 (2010) 353.
10. T. Trupke, M. A. Green, P Würfel, *J. of Appl. Phys.* 92 (2002) 4117.

11. X Wei, S Huang, Y Chen, C Guo, M Yin, W Xu, *J. Appl. Phys.* 107 (2010) 103107.
12. T. Trupke, E. Pink, R.A. Bardos, M.D. Abbott, *Appl. Phys. Lett.* 90 (2007) 93506.
13. E. Pavitra, G. Seeta Rama Raju, J. H. Oh, J. S. Yu, *New J. Chem.* 38 (2014) 3413.
14. R. Dey, V. K. Rai, *Dalton Trans.* 43 (2014) 111.
15. T. Zhou, Y. Zhang, Z. Wu, B. Chen, *J. Rare Earths* 33 (2015) 686
16. T. Trupke, A. Shalav, B.S. Richards, P. Würfel, M.A. Green, Solar Energy Materials and Solar Cells 90 (2006) 3327.
17. E. Klampaftis, B.S. Richards, Progress in Photovoltaics: Research and Applications 19 (2011) 345.
18. M.G. Debije, P.P.C. Verbunt, B.C. Rowan, B.S. Richards, T.L. Hoeks, Applied optics 47 (2008) 6763.
19. J. D. Chen, H. Guo, Z. Q. Li, H. Zhang, Y. X. Zhuang, *Opt. Mater.* 32 (2010) 998.
20. K Deng, T Gong, L Hu, X Wei, Y Chen, M Yin, *Opt. Express* 19 (2011) 1749.
21. D Chen, Y Yu, H Lin, P Huang, Z Shan, Y Wang, *Opt. Lett.* 35 (2010) 220.
22. P. Vergeer, T. J. H. Vlugt, M. H. F. Kox, M. I. Den Hertog, J. Van Der Eerden, A. Meijerink, *Phys. Rev. B* 71 (2005) 014119.
23. M. Xie, Y. Tao, Y. Huang, H. Liang, Q. Su, *Inorg. Chem.* 49 (2010) 11317.
24. Q. Y. Zhang, C. H. Yang, Z. H. Jiang, X. H. Ji., *Appl. Phys. Lett.* 90 (2007) 061914.
25. Q. Y. Zhang, G. F. Yang, Z. H. Jiang, *Appl. Phys. Lett.* 91 (2007) 051903.
26. L Xie, Y. Wang, H. Zhang, *Appl. Phys. Lett.* 94 (2009) 061905.
27. D. Q. Chen, Y. S. Wang, Y. L. Yu, P. Huang, F. Y. Weng, *J. Appl. Phys.* 104 (2008) 116105.
28. J. Chen, H. Guo, Z. Li, H. Zhang, Y. Zhuang, *Opt. Mater.* 32 (2010) 998.
29. X. Y. Huang, D. C. Yu, Q Y Zhang, *J. Appl. Phys.* 106 (2009) 113521.
30. J. Ueda , S. Tanabe, *J. of Appl. Phys.* **106** (2009) 043101.
31. J. Zhou, Y. Teng, S Ye, Y Zhuang, J Qiu, *Chem. Phys. Lett.* 486 (2010) 116.
32. E. Klampaftis, D. Ross, K.R. McIntosh, and B.S. Richards, Solar Energy Materials and Solar Cells 93 (2009) 1182.
33. A.V. Kir'yanova, V. Aboitesa, A.M. Belovolovb, M.J. Damzenc, A. Minassianc, M.I. Timoshechkinb, and M.I. Belovolovb, *J. Lumin.* 102 (2003) 715.
34. R.C. Garvie, R.H. Hannink, and R.T. Pascoe, *Nature* 258 (1975) 703.
35. N. Rakov, J.A.B. de Barbosa, R.B. Guimaraes, G.S. Maciel, *J. Alloys Compds.* 534 (2012) 32.
36. P. Wright and A.G. Evans, *Curr. Opin. Solid State Mater. Sci.* 4 (1999) 255.
37. S. Ye, Y. Katayama, S. Tanabe, J. Non-Cryst. *Solids* 357 (2011) 2268.
38. I. A. A. Terra, L. J. Borrero-Gonzalez, J. M. Carvalho, M. C. Terrile, M. C. F. C. Felinto, H. F. Brito, L. A. O. Nunes, *J. Appl. Phys.* 113 (2013) 073105.
39. J. Li, J. Zhang, Z. Hao, X. Zhang, J. Zhao, Y. Luo, *J. Appl. Phys.* 113 (2013) 223507.
40. W. Peng, S. Zou, G. Liu, Q. Xiao, J. Meng, R. Zhang, *J. Rare Earths* 29 (2011) 330.
41. J. Li, J. Zhang, Z. Hao, L. Chen, X. Zhang, Y. Luo, *Chem. Phys. Chem.* 16 (2015) 1366.
42. F. Wang, X. Liu, *Comprehensive Nanoscience and Technology*, vol. 1. London: Elsevier (2011) 607.
43. T. R. Hinklin, S. C. Rand, R. M. Laine, *Adv. Mater.* 20 (2008) 1270.
44. L. Han, C. Guo, Z. Ci, C. Wang, Y. Wang, Y. Huang, *Chem. Eng. J.* 312 (2017) 204.
45. M. Saif, N. Alsayed, A. Mbarek, M. El-Kemary, M. S. A. Abdel-Mottaleb, *J. Mol. Struct.* 1125 (2016) 763.
46. G. Chen, F. Wang, J. Yu, H. Zhang, X. Zhang, *J. Mol. Struct.* 1128 (2017) 1.

47. J. K. Cao, F. F. Hu, L. P. Chen, H. Guo, C. Duan, M. Yin, *J. Alloys Compd.* 693 (2017) 326.
48. L. Shirmane, C. Feldmann, V. Pankratov, *Phys. B Condens. Matter* 504 (2017) 80.
49. B. Wang, L. Sun, H. Ju, *Solid State Commun.* 150 (2010) 1460.
50. V. Mahalingam, F. Mangiarini, F. Vetrone, V. Venkatramu, M. Bettinelli, A. Speghini, J. A. Capobianco, *J. Phys. Chem. C* 112 (2008) 17745.
51. B. Zhou, *Nature Nanotechnol.* 10 (2015) 251.
52. M. Haase, H. Schäfer, *Angew. Chem.* International Edition. 50 (2011) 5808.
53. F. Auzel, *Chem. Rev.* 104 (2004) 139.
54. D. R. Gamelin, H. U. Güdel, *Acc. Chem. Res.* 33 (4) (2000) 235.
55. Z. Huang, X. Li, M. Mahboub, K. M. Hanson, V. M. Nichols, H. Le, M. L. Tang, C. J. Bardeen, *Nano Lett.* 15 (8) 5552.
56. M. Wu, D. N. Congreve, M. W. B. Wilson, J. Jean, N. Geva, M. Welborn, V. T. Van, V. B. Bulović, G. Moungi, *Nat. Photonics* 10 (2015) 31.
57. S. Georgescu, A. Ştefan, O. Toma, and A.M. Voiculescu, *J. Lumin.* 162 (2015) 174.
58. J. Li, J. Zhang, Z. Hao, X. Zhang, J. Zhao, Y. Luo, *Appl. Phys. Lett.* 101 (2012) 121905.
59. K.W. Krämer, D. Biner, G. Frei, H. U. Güdel, M. P. Hehlen, and S. R. Lüthi, *Chem. Mater.* 16 (2004) 1233.
60. J. Li, J. Zhang, Z. Hao, X. Zhang, J. Zhao, Y. Luo, *Chem. Phys. Chem.* 14 (2013) 4114.
61. J. Li, J. Zhang, Z. Hao, L. Chen, X. Zhang, Y. Luo, *Chem. Phys. Chem.* 16 (2015) 1366.
62. J. Li, J. Zhang, X. Zhang, Z. Hao, Y. Luo, *J. Alloy. Compd.* 583 (2014) 96.
63. Y. Shimomura, T. Kurushima, N. Kijima, *J. Electrochem. Soc.* 154 (2007) J234.
64. Z. D. Hao, J. H. Zhang, X. Zhang, S. Z. Lü, X. J. Wang, *J. Electrochem. Soc.* 156 (3) (2009) H193.
65. Z. Hao, J. Zhang, X. Zhang, X. Wang, *Opt. Mater.* 33 (2011) 355.
66. J. R. Carter, R. S. Feigelson, *J. Am. Ceram. Soc.* 47 (1964) 141.
67. R. D. Shannon, *Acta Crystallogr. A* 32 (1976) 751.
68. R. Gaume, B. Viana, J. Derouet, D. Vivien, *Opt. Mater.* 22 (2003) 107.
69. K. Yu. A. V. Voron'ko, K. N. Malov, P. A. Nishchev, A. A. Ryabochkina, S. Sobol, N. Ushakov, *Opt. Spectrosc.* 102 (2007) 722.
70. F. Cornacchia, A. Toncelli, M. Tonelli, E. Favilla, K. A. Subbotin, V. A. Smirnov, D. A. Lis, E. V. Zharikov, *J. Appl. Phys.* 101 (2007) 123113.
71. S. Georgescu, A. M. Voiculescu, C. Matei, A. G. Stefan, O. Toma, *Phys. B* 413 (2013) 55.
72. R. Yan, Y. Li, *Adv. Funct. Mater.* 15 (2005) 763.
73. L. G. van Uitert, L. F. Johnson, *J. Chem. Phys.* 44 (1966) 3514.
74. R. V. Bakaradze, G. M. Zverev, G. Ya Kolodnii, G. P. Kuznetsova, A. M. Onischenko, *Solid State Phys.* 9 (1967) 939.
75. A. Ştefan, O. Toma, Ş. Georgescu, *J. Lumin.* 180 (2016) 376.
76. Q. Cao, D. Peng, H. Zou, J. Li, X. Wang, X. Yao, *J. Adv. Dielectr.*, 04 (2014) 1450018.
77. R. Adhikari, J. Choi, R. Narro-García, E. De la Rosa, T. Sekino, S. W. Lee, *J. Solid State Chem.* 216 (2014) 36.
78. T. Wei, C. Z. Zhao, Q. J. Zhou, Z. P. Li, Y. Q. Wang, L. S. Zhang, *Opt. Mater.* 36 (2014) 1209.
79. Z. W. L. X. L. Yang, S. G. Xiao, *Spectrosc. Spectr. Anal.* 22 (2002) 357.
80. Y. Li, X. Wei, M. Yin, *J. Alloy. Compd.* 509 (2011) 9865.

81. M. X. Façanha, J. P. C. do Nascimento, M. A. S. Silva, M. C. C. Filho, A. N. L. Marques, A. G. Pinheiro, A. S. B. Sombra, *J. Lumin.* 83 (2017) 102.
82. Z. S. Chen, T. F. Chen, W. P. Gong, W. Y. Xu, D. Y. Wang, Q. K. Wang, *J. Am. Ceram. Soc.* 96 (2013) 1857.
83. H. L. Han, L. W. Yang, Y. X. Liu, Y. Y. Zhang, Q. B. Yang, *Opt. Mater.* 31 (2008) 338.
84. T. Li, C. F. Guo, Y. R. Wu, L. Li, J. H. Jeong, *J. Alloys Compd.* 540 (2012) 107.
85. X. X. Luo, W. H. Cao, *J. Mater. Res.* 23 (2008) 2078.
86. O. A. Lopez, J. McKittrick, L. E. Shea, *J. Lumin.* 71 (1997) 1.
87. G. Y. Chen, H. C. Liu, H. J. Liang, G. Somesfalean, Z. G. Zhang, *J. Phys. Chem. C* 112 (2008) 12030.
88. B. P. Singh, A. K. Parchur, R. S. Ningthoujam, P. V. Ramakrishna, S. Singh, P. Singh, *Phys. Chem. Chem. Phys.* 16 (2014) 22665.
89. H. J. Liang, Y. D. Zheng, G. Y. Chen, L. Wu, Z. G. Zhang, W. W. Cao, *J. Alloys Compd.* 509 (2011) 409.
90. L. Jiang, S. Xiao, X. Yang, J. Ding, K. Dong, *Appl. Phys. B* 107 (2012) 477.
91. L. Jiang, S. Xiao, X. Yang, J. Ding, and K. Dong, *Appl. Phys. B* 107 (2012) 477.
92. T. Li, C.F. Guo, Y.R. Wu, L. Li, and J.H. Jeong, *J. Alloys Compds.* 540 (2012) 107.
93. M. Pokhrel, G.A. Kumar, D.K. Sardar, *Mater. Lett.* 99 (2013) 86.
94. X. Q. Chen, Z. K. Liu, Q. Sun, M. Ye, F. P. Wang, *Opt. Comm.* 284 (2011) 2046.
95. G.S. Qin, W.P. Qin, S.H. Huang, C.F. Wu, D. Zhao, B.J. Chen, S.Z. Lu, E. Shulin, *J. Appl. Phys.* 92 (2002) 6936.
96. M. Pollnau, D. R. Gamelin, S. R. Luthi, H. U. Gudel, *Phys. Rev. B* 61 (2000) 3337.
97. Y. Guo, D. Wang, F. Wang, *Opt. Mater.* 42 (2015) 390.
98. M. Haase, H. Schafer, *Angew. Chem. Int. Ed.* 50 (2011) 5808.
99. L.N. Guo, Y.H. Wang, J. Zhang, Y.Z. Wang, P.Y. Dong, *Nanoscale Res. Lett.* 7 (2012) 1.
100. S. Singh, J.E. Geusic, *Phys. Rev. Lett.* 17:(16) (1966) 865.
101. X. Q. Chen, Z. K. Liu, Q. Sun, M. Ye, F. P. Wang, *Opt. Comm.* 284 (2011) 2046.
102. W.P. Zhang, P.B. Xie, C.K. Duan, K. Yan, M. Yin, L.R. Lou, S.D. Xia, and J.C. Krupa, *Chem. Phys. Lett.* 292 (1998) 133.
103. W. Xu, X.Y. Gao, L.J. Zheng, Z.G. Zhang, and W.W. Cao, *Opt. Express* 20:(16) (2012) 18127.
104. Y. Ledemi, M.E. Amraoui, J.L. Ferrari, P. Fortin, S.J.L. Ribeiro, and Y. Messaddeq, *J. Am. Ceram. Soc.* 96 (2013) 825.
105. Y. Teng, K. Sharafudeen, S. F. Zhou, J. R. Qiu, *J. Ceram. Soc. Jpn.* 120 (1407) (2012) 458.
106. I. W. Donald, P. M. Mallinson, B. L. Metcalfe, L. A. Gerrard, J. A. Fernie, *J. Mater. Sci.* 46 (7) (2011) 1975.
107. T. Hoche, *J. Mater. Sci.* 45 (2010) 3683.
108. J. R. Qiu, *J. Ceram. Soc. Jpn.* 116 (2008) 593.
109. W. Holand, V. Rheinberger, M. Schweiger, *Adv. Eng. Mater.* 3 (2001) 768.
110. G. Partridge, *Glass Technol.* 35 (1994) 116.
111. G. Partridge, *Glass Technol.* 35 (1994) 171.
112. G. Partridge, S. V. Phillips, *Glass Technol.* 35 (1991) 82.
113. P. A. Tanner, C. K. Duan, *Coord. Chem. Rev.* 254 (2010) 3026.
114. L. N. Guo, Y. H. Wang, J. Zhang, Y. Z. Wang, P. Y. Dong, *Nanoscale Res. Lett.* 7 (2012) 1.

115. W. Xu, X. Y. Gao, L. J. Zheng, Z. G. Zhang, W. W. Cao, *Opt. Express* 20 (2012) 18127.
116. S. Jiang, H. Guo, X. Wei, C. Duan, M. Yin, *J. Lumin.* 152 (2014) 195.
117. W. J. Zhang, Q. J. Chen, Q. Qian, Q. Y. Zhang, *J. Am. Ceram. Soc.* 95 (2012) 663.
118. M. Haase, H. Schäfer, *Angew. Chem. Int. Ed.* 50 (2011) 5808.
119. Q. Shi, J. Zhang, C. Cai, L. Cong, T. Wang, *Mater. Sci. Eng. B* 149 (2008) 82.
120. J. Liu, X. Wang, T. Xuan, H. Li, Z. Sun, *J. Alloy. Compd.* 593 (2014) 128.
121. X.Y. Liu, Y.L. Wei, R.F. Wei, J.W. Yang, and H. Guo, *J. Am. Ceram. Soc.* 96:(3) (2013) 798.
122. X. Y. Huang, S. Y. Han, W. Huang, X. G. Liu, *Chem. Soc. Rev.* 42 (2013) 173.
123. X. Y. Huang, J. X. Wang, D. C. Yu, S. Ye, Q. Y. Zhang, X. W. Sun, *J. Appl. Phys.* 109 (2011) 113526.
124. W. He, T. S. Atabaev, H. K. Kim, Y. H. Hwang, *J. Phys. Chem. C* 117 (2013) 17894.
125. M. J. Lim, Y. N. Ko, Y. Chan Kang, K. Y. Jung, *RSC Adv.* 4 (2014) 10039.
126. J. Dewalque, R. Cloots, F. Mathis, O. Dubreuil, N. Krins, C. Henrist, *J. Mater. Chem.* 21 (2011) 7356.
127. S. G. Shin, K. H. Kim, C. W. Bark, H. W. Choi, *J. Korean Phys. Soc* 65 (2014) 387.
128. C. W. Kim, D. K. Kim, W. J. Shin, M. J. Choi, Y. S. Kang, Y. S. Kang, *Nano Energy* 13 (2015) 573.
129. X. Huang, S. Han, W. Huang, X. Liu, *Chem. Soc. Rev.* 42 (2013) 173.
130. Y. Tai, G. Zheng, H. Wang, J. Bai, *J. Solid State Chem.* 226 (2015) 250.
131. C. K. Jørgensen, *Prog. Inorg. Chem.* 12 (1970) 101.
132. H. Lin, D. Chen, Y. Yu, *J. Alloys Compd.* 509 (2011) 3363.
133. A. S. Vanetsev, Y. V. Orlovskii, V. Nagirnyi, I. Sildos, L. Puust, K. Kaldvee, M. Yin, X. T. Wei, V. N. Makhov, *Radiat. Meas.* 90 (2016) 329.
134. J. Liao, D. Zhou, S. Liu, H. R. Wen, X. Qiu, J. Chen. *Physica B* 436 (2014) 59.
135. C. H. Chiu, M. F. Wang, C. S. Lee, T. M. Chen, *J. Solid State Chem.* 180 (2007) 619.
136. R. T. Wegh, H. Donker, K. D. Oskam, A. Meijerink, *Science* 283 (1999) 663.
137. C. Feldmann, T. Jüstel, C. R. Ronda, D. U. Wiechert, *J. Lumin.* 92 (2001) 245.
138. D. Chen, Y. Yu, Y. Wang, P. Huang, F. Weng, *J. Phys. Chem. C,* 113 (2009) 6406.

Part IV

Efficient and Eco-Friendly Technology

6

Photovoltaic Cell Response and Other Solar Cell Materials

6.1 Solar Photovoltaic Technology

Photovoltaic (PV) technology is a promising and clean source of energy generation for the 21st century, because it has the ability to reduce pollution and also helps to reduce the level of harmful pollutants, which are continuously growing due to the increase in the use of traditional polluting technology for energy generation, such as coal power and oil-based plants, and also the use of agricultural waste in power generation. Recently, solar cells based on silicon and other inorganic materials have been commercialized and are now applicable everywhere for household and some small-scale industrial purpose. Currently, the use of solar cells for energy applications is increasing due to their global advantages. But to date, solar cell technology does not fully cover rural and urban areas due to its high installation cost; it is not affordable for everyone because of its high industrial processing cost, which is a major concern, and to date there is no substitute available to replace Si-based solar cells. In underdeveloped countries, governments are providing much support to develop and use renewable and environmentally friendly sources of energy for sustainable development to avoid issues related to global warming. Therefore, it is a big challenge for researchers to develop environmentally friendly technology. We need to develop materials that are cheaper, with better efficiencies and low fabrication cost. To date, solar cells are on a limited scale for residential purposes. We have to develop vacuum-free fabrication methods for Si cells, which is a leading way to develop PV technology. We know that most thin film cells are fabricated using casting and printing methods. Thus, solution-processed thin film solar cells have potential application on a commercial scale with low fabrication costs. With the increasing demand for new, reliable, clean, equitable substitutes for finite energy resources, solar energy is the best substitute in the current century to avoid the use of traditional fuel-based technology, which causes serious environmental problems and also, when used in excess, harms to human health and has several impacts on social progress [1]. Currently, one of the greatest challenges in PV technology is how to increase the current cell efficiency of silicon solar cell by capturing the maximum solar radiation incident on them and generating clean, affordable solar electricity [2]. As discussed earlier in the case of the first generation, the most developed PV solar cells fully dominate the market and are used all over the world. The cost of second-generation devices is greatly reduced while achieving better cell efficiency. And today we know that, in the third generation of PV cells, effort has been continuously ongoing toward the use of luminescent materials, solar concentrators, and organic-based PV cells to develop the PV technology. As yet, they are not widely commercialized [3–5]. Further, dye-sensitized nanocrystalline PV cells are an interesting candidate due to their low fabrication cost, variety of colors, easy manufacturing process, and clean, eco-friendly, and relatively high power conversion efficiency [6–14]. The next section of this chapter gives a detailed account of the current efficiency of the Si solar cell and an ideal way to enhance its efficiency by using a thin film spectral converter for the development of PV solar technology.

6.1.1 Efficiency of Si-Solar Cells

Figure 6.1 shows the spectral energy distribution of an AM 1.5 G solar cell reader indicating the fraction of light absorbed by a crystalline silicon (c-Si) solar cell. This means that the absorption range of Si solar cells matches a spectral wavelength of sunlight, and other wavelengths, indicated by the blue color, cannot be absorbed by Si solar cells; that is, there is an unused component of light coming from the sun. To cover wavelengths of incident light other than the absorption edge of the Si cell is only possible by using the downshifting, downconversion, and upconversion mechanisms with the help of an active layer. These techniques find more advantages in the third generation of PVs to enhance the light conversion efficiency of the cell. Hence, from Figures 6.1 and 6.2, it is seen that the sunlight falling in the ultraviolet to visible spectral range cannot be used by the solar cell, and the wavelength of light corresponding

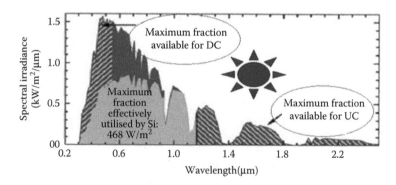

FIGURE 6.1
Plot of wavelength versus spectral irradiance indicating the amount of light absorbed by a c-Silicon cell.

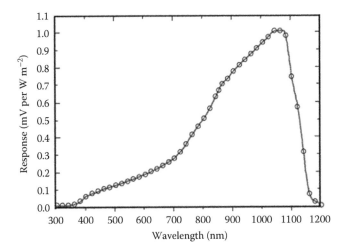

FIGURE 6.2
Spectrum showing the response of a c-Si solar cell.

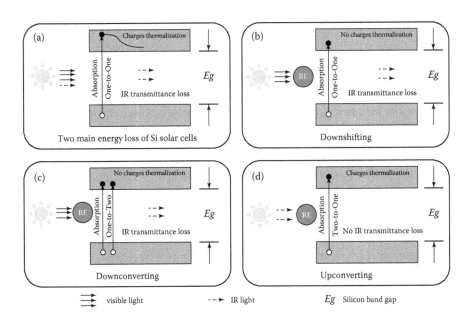

FIGURE 6.3
Representation of (a) loss observed in silicon solar cell and the process used to modify the efficiency of the cell by using (b) downshifting, (c) downconversion, and (d) upconversion mechanism.

to the near infrared region of the solar spectrum can be caught by the Si cell by converting it into the band gap range of Si, which is approximately 1.12 eV (shorter wavelength). In the Si solar cell, approximately 70% energy losses occur due to the spectral mismatch. Another factor that is responsible for the loss is the thermalization of carriers. Figure 6.3a shows the two energy loss mechanisms occurring in the light conversion process. The first is due to the thermalization process and the other to a transmitted component of infrared light, which does not participate in the process of photoconversion. For this reason, the existing efficiency of c-Si solar cells can reach 15%–20%, and the theoretical value reported in the literature is 30% [16–18].

6.1.2 Efficiency with Coating of Upconversion/ Downconversion (UC/DC) Phosphor Layer

Figure 6.3b, as mentioned in the previous section, shows the conversion of high-energy photons (UV-visible range) into low-energy photons (visible–near infrared region), which are then used by Si solar cells by the process of a down-shifting mechanism having maximal quantum efficiency (100%). However, it is well known that the thermalization of the charge carriers will greatly affect the efficiency of Si solar cells. The power output of the cell is also enhanced, but it does not exceed the reported value for efficiency, that is, 30%. Figure 6.3c shows the ultraviolet–green region of the high-energy photon converted into two near infrared photons of low energy, which are absorbed by the Si solar cell by the process of downshifting, with an efficiency of about 200%.

In this process, a number of photons that correspond to the ultraviolet–visible region of the spectrum participate in the light conversion process and enhance the efficiency of the cell, thereby decreasing the thermalization loss due to the spectral matching between incident near infrared photons and the absorption region of the Si solar cell. In this case, the efficiency of the Si solar cell reached 40%, whereas the low-energy photons correspond to the infrared region were not used, but they can be converted into high-energy photons of the visible range before being absorbed by the Si solar cell by means of the upconversion process; in this case, the loss observed due to transmission will be greatly reduced, as indicated in Figure 6.3d. Therefore to solve such problems associated with Si solar cells, rare earth–doped inorganic lumines-cent materials have found potential application as wavelength converters to enhance the conversion efficiency. In the past few years, many research groups have been busy finding new and high efficient luminescent materials as spec-tral converters using the different combination of rare earth ions, such as double-doped and triple-doped ions, and so on. Among them, double-doped and triple-doped rare earth ions for the spectral converter, downshifting, and downconversion play an important role in enhancing the efficiency of solar cells. In the case of inorganic luminescent materials, compounds with better physical and chemical properties and also good thermal stability, quantum efficiency and so on are always in demand. Rare earth ions have the ability to act as sensitizers to use the high-energy photon of incident sunlight and also as activators that convert the sun's high-energy photons into photons of longer wavelength, which fulfill the conditions of energy enhancement in the Si solar cell. Therefore, the focus has been on the development of rare earth ion–based solar converter materials due to the allowed and forbidden nature of rare earth transitions. Many reports have been published on the development of rare earth–based c-Si solar cells. Thus, from the reported literature, it is seen that the use of a UC/DC phosphor coating on the Si solar cell is an ideal way to enhance the light conversion efficiency of the Si solar cell. Also, efforts have been made to reduce the fabrication cost of such solar cells in the near future. The next paragraph provides a discussion on recently published work on the downshifting $Gd_2O_2S:Eu^{3+}$ phosphor.

Recently, Chen et al. [92] reported downshifting in $Gd_2O_2S:Eu^{3+}$ phosphor combined with polyvinylpyrrolidone (PVP) composite film and explored the practical concept of the rare earth–activated phosphor as a spectral converter for a Si solar cell, which is depicted in Figure 6.4a. The prepared

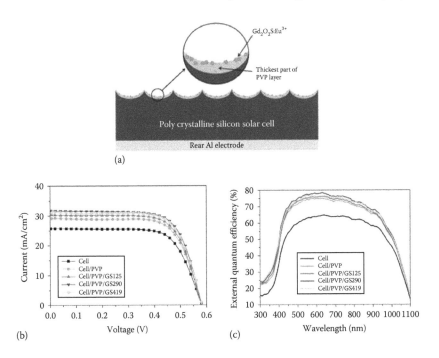

FIGURE 6.4

(a) Schematic of rare earth spectral conversion using $Gd_2O_2S:Eu^{3+}$ phosphor–PVP film for Si solar cells. (b) Current-voltage characteristics of the rare earth crystalline Si solar cell, showing the cell response before and after coating PVP–$Gd_2O_2S:Eu^{3+}$ phosphor film. (c) Graph showing the external quantum efficiency versus wavelength for the Si solar cell before and after coating with PVP and PVP–$Gd_2O_2S:Eu^{3+}$ phosphor. (Reprinted from Hung, W. B. and Chen T. M. Efficiency enhancement of silicon solar cells through a downshifting and antireflective oxysulfide phosphor layer. *Sol. Energy Mater. Sol. Cells,* 133, 39, 2015, Copyright. with permission from Elsevier.)

submicron-sized $Gd_2O_2S:Eu^{3+}$ phosphor particles, with a good crystallite size, show the characteristic emission of the Eu^{3+} ion. The photoluminescence excitation extends over a broad range from 250 to 400 nm, whereas the emission corresponds to the red region of the spectrum centered on 624 nm, assigned to the f–f transition of the Eu^{3+} ion. The authors reported that the photoluminescence intensity for GS_{125}, GS_{290}, and GS_{419} was equal to 30.7%, 35.7%, and 45.8%, respectively, with respect to $Y_2O_2S:Eu^{3+}$ phosphor [92]. And lastly, luminescent rare earth–based spectral converters will be studied from the practical point of view. Further considering the upcoming challenges in Si solar cells, rare earth ions having better conversion efficiency act as spectral converters, and these cells are called *REC-Si solar cells*. In the last few years, rare earth ions doped into near-infrared downconversion and downshifting inorganic-based phosphors, which are generally crystalline materials or thin films, have attracted special interest, and in the past few years there are many works that have been published on rare earth solar light converters. This may give strong support for proving whether or not they really have the ability to enhance the cell response of existing Si solar cells in light energy conversion. Today, new experimental evidence is observed for the downshifting phosphor layer, which is the combination of microcrystalline phosphor with PVP composite film spread on the Si solar cell. Figure 6.4b shows the current-voltage characteristics of a rare earth crystalline cell coated with a combination of enhancement PVP and $PVP/Gd_2O_2S:Eu^{3+}$ phosphor layers. The cell response reached 10.44%, the current capacity (density) was found to be 25.76% mA/cm^2, and the corresponding value of open-circuit voltage (Voc) was reported as 0.5790 V with fill factor (FF) = 69.9 after the layer of PVP and $Gd_2O_2S:Eu^{3+}$/PVP phosphor was applied on the top surface of the Si solar cell. Therefore, the maximum increase in short circuit density (Jsc) was reported at 6 mA/cm^2, which corresponds to a 23% increase in value as compared with the bare cell; this can be achieved only by combining the PVP/GS290 as an active layer on the Si solar cell. Here, the increase in efficiency was reported to be 8.9% with a coating of GS290/PVP rather than PVP. This shows that positive results are obtained with an increase in light absorption and cell response. Here the reflected light is dominated by applying $G_2O_2S:Eu^{3+}$/PVP layer especially by coating with PVP/GS290 layer. The phenomenon of light emission trapping and enhancement in light intensity is obtained by scattering of light from the $Gd_2O_2S:Eu^{3+}$ phosphor particles; hence, it reduced the reflectance. Figure 6.4c, shows a graph of the effective quantum efficiency (EQE) versus wavelength of the Si solar cell. It is concluded that the enhancement in light conversion efficiency occurs not only due to the antireflection coating layer but also due to the downshifting process of $Gd_2O_2S:Eu^{3+}$ phosphor particles [19]. To protect the c-Si solar cell from environmental moisture, it is encapsulated into a module to avoid any damage that might affect the cell's efficiency. Therefore, for a modular structure of this kind, special attention has been given to avoiding any loss in absorption and reflection of light from the front glass surface [20]. Many researchers are currently busy with the development of rare earth c-Si solar cells via the process of the downshifting mechanism. However, this is not enough from the practical point of view; there is an urgent need to develop highly efficient inorganic materials that have better conversion efficiency.

6.2 Some Reported Work

Recently, Li et al. [20], reported a downshifting in Eu^{3+} complexes and explored their use in a rare earth ion–based solar spectral converter for a converted Si solar cell [20]. They studied the EQE of monocrystalline silicon PV modules with and without polyvinyl acetate (PVA) film and also coated with combination PVA films doped with different complexes (numbered from 1 to 5). Along with this, they studied the I-V characteristics of an uncoated (c-Si) PV module and c-Si PV module coated with PVA film doped with Complex 5. Photoluminescence emission spectra of the Eu^{3+} complex ion similar to those of $Gd_2O_2S:Eu^{3+}$ phosphor were observed. After applying a thin PVA–complex layer on the Si solar cell, they reported that Jsc values increased from 35.67 to 36.38 mA/cm²; however, the conversion efficiency (η) of the cell increased in the range of 16.05%–16.37%. Hence, the maximum efficiency was reported to be above 50% in the wavelength range of 300–400 nm. Also, Figure 6.5(b) depicted the enhancement in the EQE of a c-Si module by coating with a PVA–complex layer (Complexes 1, 2, 3, 4, and 5); this was achieved due to light conversion efficiency of complexes. In this case, the authors predicted that minimal loss would occur due to reflection [20]. From the above-mentioned results, it is concluded that the downshifting process via a Eu^{3+}-doped complex plays an important role in improving the spectral response of the Si solar cell. However, such a process using a downshifting mechanism is not an ideal technique for light conversion in improving the spectral response of the solar cell. By using this concept, we can cover only a small fraction of the UV component of the solar radiation. The downshifting layer is important to convert ultraviolet into visible photons, but it does not match the absorption range of the Si solar cell, so that this creates a larger gap in enhancing the efficiency of the Si solar cell [21].

The rare earth–doped near-infrared downconversion or downshifting materials have the following challenges.

1. To date, there are very few highly efficient downconversion or downshifting materials that have the best EQE. And another important factor is the better wavelength range and nature of absorption and emission spectra than those DS materials that are proposed for the solar cell application to enhance the light conversion efficiency. Hence, it will be a future challenge in the field of PV to develop new and advanced materials with high EQE [22].

2. There is also a need to focus on the film transparency of near-infrared downconversion materials. We have to find new and advanced glass or composite materials for polymer films with better transparency and optical properties [22].

3. Another challenge associated with PV devices is how to apply the spectral converting layer. Therefore, to resolved this issue, a concept experiment has been performed on a Si solar cell module based on near infrared downshifting/downconversion materials [22].

6.3 Types of Solar Cell

A PV cell is an electrical device that works on the basis of the PV effect. Different types of solar cell are currently available. The following is a list of solar cells and the raw materials used to fabricate them.

Type of Solar Cell	Materials Used
Amorphous ⎫	
Crystalline ⎬ Si solar cell	Silicon
Polycrystalline ⎭	
Biohybrid solar cell	Organic and inorganic materials
CdTe solar cell	Inorganic materials
GaAs solar cell	Semiconductor materials
Hybrid solar cell	Organic and inorganic materials
Perovskite solar cell	Organic and inorganic halide materials
Plastic solar cell (also called *organic solar cell*)	Conducting polymer or organic materials
Quantum dot solar cell	Silicon, copper, indium
	Gallium selenide bulk materials

Among these types, Si solar cells are the most commonly used all over the world. Therefore, from this list, only three cells will be selected for a brief discussion due to their cell efficiency and easy fabrication process. These are the organic, thin film, and perovskite solar cells. Figure 6.5 presents the efficiency of different types of solar cell as mentioned earlier. Depending on the light conversion efficiency, any one of these may be a future substitute for the Si-based solar cell [23–36].

6.3.1 Organic Solar Cell

Also called a *plastic solar cell*, this is based on an organic device. This type of cell is made using a conducting polymer or an organic compound [37]. Such polymer film is transparent in nature and it absorbs the light incident on

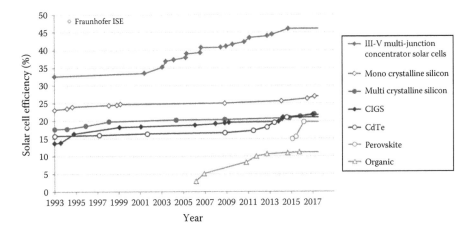

FIGURE 6.5
Cell efficiencies of different types of solar cell. (From the photovoltaics report, Fraunhofer Institute for Solar Energy Systems, Freiburg, 12 July 2017).

it and generates the electric current. Hence, such a device is called a polymer solar cell. This electrical device or cell is produced by simple solution-based techniques. Therefore, the mass scale production cost to fabricate an organic cell is also lower [38]. Considering the other advantages, such as the cost of the cell, efficiency, and so on, an organic cell is a promising candidate for PV application. The basic properties of polymer materials can change by varying their length and functional group, which affect the energy gap and conduction of materials. Organic materials can absorb most of the incident light even in small materials due to their high absorption coefficient, which is of nanometer size. This cell faces drawbacks such as low quantum efficiency and lower strength compared with the Si solar cell. However, a cell made up of polymer materials is cheap to fabricate and more flexible to modify on a small scale. It has less adverse environmental impact. Because of its transparent nature, it may be applicable to window, wall, and other electronic equipment. Figure 6.6 shows the schematic of a polymer solar cell. The main drawback of this cell is that it offers only one-third of the efficiency as compared with hard materials, and it suffers from degradation [39] and thermal stability [40] despite its low-cost fabrication technique [41]. Recently, its efficiency has reached 10% via a tandem structure [42], which has made polymer cells an interesting field in PV technology. This cell has a conjugate structure, and it is formed by the combination of single and double bonds covalently bonded with the carbon atom. Here, the electron Pz orbitals are delocalized and form π and π^* bonding and antibonding orbitals; the π orbitals are highly occupied (HOMO) and the π^* orbitals are unoccupied orbitals (LUMO). Therefore, in organic devices, HOMO and LUMO act as the valence and conduction bands, respectively. The energy band gap of these materials lies between 1 and 4 eV [43]; thus, the incident light, having energy larger than the band gap

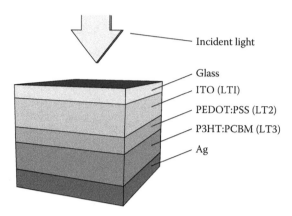

Incident light

Glass
ITO (LTI)
PEDOT:PSS (LT2)
P3HT:PCBM (LT3)
Ag

FIGURE 6.6
Fabrication of thin-film organic solar cell (OSC).

value of the material, can be easily absorbed, resulting in low conversion efficiency and lower voltage. In recent years, bulk organic solar cell materials based on poly(3-hexylthiophene) (P3HT) and phenyl-C61-butyric acid methyl ester (PCBM), or $P_3HT:PCBM$, have proved to be efficient materials and have gained much interest due to their large absorption coefficient [44]. Many researchers are attempting to develop highly absorptive organic solar cell materials to design and fabricate a solar cell that enables resonant oscillation of electron excitation; this has been explored to increase the total absorption of the cell [45–48, 49–61]. To date, many plasmonic solar cell strategies have been reported in the literature. In general, there are three applicable ways to increase the optical absorption in the organic solar cell. The first way is to spread a layer of metal nanoparticles on the top of the absorbing materials [45, 50, 49, 60]. The second is to integrate metallic nanoparticles into the active materials [48, 49], and the third is by placing random metallic nanoparticles or a metallic structure on the back surface contact to use polaritons at the metal interface [60]. In the case of $P_3HT:PCBM$ solar cells, the optical absorption is enhanced by using a mixture of metal nanoparticles and absorbing materials. But as yet, no report is available on the incorporation of metallic nanoparticles on the back contact surface of the organic solar cell. Figure 6.7 shows a schematic representation of an organic thin film solar cell. It is fabricated by the combination of poly (3,4-ethylenedioxythiophene), poly (styrenesulfonate), poly (3-hexylthiophene) 6:6] -phenyl-C61-butyric acid/silve/indium tin oxide, or $PEDOT:PSS/P_3HT:PCBM/AG/ITO/glass$. The main characteristic features that distinguish organic PVs from inorganic PVs is that the inorganic materials create free charge carriers after the absorption of a photon, whereas in the case of organic materials, an excited pair or exciton is generated, from which the charge carrier still needs to be dissociated. The light conversion efficiency of the organic solar cell is low in the case of thin film formation when compared with inorganic materials. This occurs due to the balance between

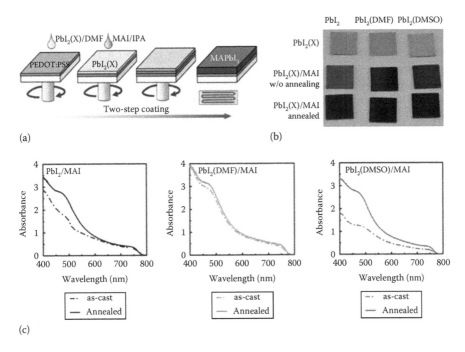

FIGURE 6.7
(a) Perovskite film fabrication by two-step coating method. (b) Formation of PbI_2 or $PbI_2(X)$ complexes and MAI film. (c) Absorption spectra of different types of perovskite film. (Reprinted from *Sol. Energy Mater. Sol. Cells,* 155, Fua, W., et al., 331, Copyright (2016), with permission from Elsevier.)

photon absorption and exciton diffusion. Hence, the photogenerated exciton is diffused in a limited range. Thus, the absorption layer required is thinner. Therefore, the optical absorption, in this case, is not sufficiently strong. Hence, the low light conversion efficiency becomes one more challenge due to the very thin active layer. However, today, there are more promising reports available on enhancing the efficiency of organic solar cell materials, which may help to fabricate highly efficient organic solar cells.

6.3.2 Perovskite Solar Cell

The perovskite solar cell is also a promising field in PV technology. Currently, it is a hot topic for researchers working in PV. One major challenge all over the world is to find advanced and highly efficient materials for a solar cell. The perovskite cell is made using a combination of organic–inorganic hybrid materials; generally, it contains a halide group. Also, perovskite materials have a few characteristic features, such as high absorption coefficient, high carrier mobility, and a low-cost device fabrication process. So, in the near future, they may be promising solar cell materials to resolve all the issues in PV technology [61]. The conversion efficiency of perovskite materials reached 4% in 2009 [62]. Therefore, considering its light weight and cheaper materials, it can replace existing solar cell devices [61–75]. The PV properties of the perovskite depend on the quality of the film, including its morphology, grain size, and grain boundary, and also on its crystallization process [76–80], in which the crystallization kinetics play an important role and the film processing parameters are varied with the help of reaction temperature [81–85], rate and time of spin coating [86–88], atmospheric conditions, the use of different solvents, and so on [89–97]. The kinetics of crystal growth in perovskite depends on inorganic and organic salts and affects the film morphology and performance of the cell [98–104]. Different deposition techniques are used for device fabrication, such as physical vapor deposition and one-step deposition [105, 106]. Other techniques have also been used, such as two-step solution deposition and the vapor-assisted reaction process [107, 108]. Liu et al. [109] reported a vapor-deposited thin film having good crystalline nature, of the order of hundreds of nanometers in size, achieving power efficiency above 15%. However, Jeon et al. [110, 111] reported one-step deposition techniques using a mixture of γ-butyrolactone and dimethyl sulfoxide to dissolve methylammonium iodide and a halide such as lead iodide by drop casting toluene application and heat treatment to achieve crystallinity. So, a uniform and polycrystalline film is obtained by applying chlorobenzene to the film for up to 6 s, without the need for heat treatment, and this film enhances the device efficiency up to 13.9% [112]. Also, a thermal annealing treatment is provided to increase the grain size and to improve the crystallinity of materials to the same extent. However, if the annealing temperature is high for a short time or low for a longer time, it may decompose the film and hence, be responsible for decreasing the device efficiency [113]. Thus, we can conclude that perovskite (organic–inorganic) materials are better materials for acting as absorbers in solar cells. Today, many research organizations are working on enhancing the efficiency of $CH_3NH_3PbX_3$ (where X=Cl, Br, I) solar cells. There is one other technique available to fabricate a sandwich-type absorber to obtain a high absorption coefficient and high carrier mobility. Also, Burschka and co-workers reported the fabrication of a highly power-efficient and highly stable cell by the two-step method [114]. Seo et al. [115], discuss the fabrication of a $CH_3NH_3PbI_3$/PbS high-performance quantum dot solar cell, whereas Geng et al. theoretically investigated the electronic, structural, and optical properties of $CH_3NH_3PI_3$ [116]. Therefore, based on the work cited here, it is seen that most of the work focused on halogen trimethylamine and iodine lead methylamine. We know that lead is poisonous and harmful to growth and development; also, tin consists of Sn^{4+} due to the oxidation process and shows metal-like behavior in a semiconductor, thereby reducing the PV effect [117]. Thus, today, materials with high absorption coefficient as compared with the above-discussed materials are always in demand for developing a perovskite-based solar cell.

Miyasaka et al. (2009) were the first to introduces the use of perovskite for a solar cell. The fabricated cell was fairly unstable, and power conversion efficiency of up to 3.8% achieved [112]. In 2013, with further invention using the planar thin film structure, cell efficiency of over 15% was reported [118]. The maximum efficiency of a perovskite cell was reported by a Korean researcher as 20.1% [119]. Perovskite materials have many advantages over the silicon cell. The device processing cost is lower, and film formation techniques are simple, as shown in Figure 6.7 [120, 121]. Also, the band gap is larger than for silicon. The material is usually applied on the top surface of the cell due to its transparency to typical solar absorber wavelengths. Therefore, this cell has the potential to cover a greater part of the solar radiation and obtain better efficiency than other solar cell devices. But at present, this type of cell faces many challenges before it can be applied on a widespread level. The first challenge is that the materials degrade quickly in water [122]. In addition to this, the efficiency of this cell is relatively high and uncertain. Another challenge is discussed by Bi et al. [123]: the actual performance of the perovskite cell in long-term practical applications is still unclear.

6.3.3 CdTe/CdS Solar Cell

Another class of solar cell that uses inorganic materials is the thin film solar cell. It shows a rapid increase in cell efficiency, up to 21% [124]. The thin layer of the cell is effectively used for energy conversion in the thin film solar cell. Thin films show better optoelectronic properties, such as band gap, high absorption coefficient, and better thermal stability [125]. This makes them interesting candidates for PV technology. Many researchers are busy developing and identifying the processes and electronic properties of materials. So that they have reported the efficiency of the cell upto 10% in 1972 but in 2016 this is reported upto 22.1% [126]. The CdTe thin film solar cell is grown on a glass substrate with a transparent conductive oxide (TCO), and another layer of oxide, called a *buffer layer*, is deposited on the TCO [127]. Usually, the CdTe layer is deposited on a buffer layer having energy greater than the band gap (2.4 eV). Some of the absorbed light is wasted due to carriers photogenerated in the CdTe not being collected [128]. The operating voltage of the cell degrades if the CdS layer is too thin. Therefore, the thickness of the CdTe film optimizes efficiency and must balance the IV at maximum power output. The low value of achievable output circuit voltage affects the cell's efficiency. This parameter depends on dopant density and minority charge carriers, which play a crucial role in crystallographic defect CdTe–based cell efficiency. Besides this, grain boundaries are also among the most prominent defects in the CdTe absorption layer and act as a diffusion pathway for the dopant. Thus, the CdTe absorber layer in the thin film solar cell is grown by two different configurations: superstrate and substrate. The two main significant features of the substrate configuration as compared with superstate are first, that it is deposited on the nontransparent and flexible substrate, and second, that it is used to control the p–n junction properties due to its stacking sequence. The basic advantage of the CdTe solar cell over the Si solar cell is that the production capacity of CdTe thin film cells

TABLE 6.1

Comparison between Crystalline and Thin Film Solar Cells

PV Cell Technology	Silicon Solar Cell	Thin Film Solar Cell
Types of solar cell	• c-Si • pc-Si	• Amorphous silicon solar cell • CdTe solar cell • Copper indium gallium selenide cell
Voltage rating	80%–85%	72%–78%
Temperature coefficients	Higher	Lower
Ge current-voltage fill factor	73%–82%	60%–68%
Efficiency of the cell	13%–19%	4%–12%

increased significantly in 2007, whereas the cost of crystalline Si modules fell rapidly [129]. However, the cell efficiency of CdTe and Si solar cell modules has been rapidly increasing, whereas the cost difference between the cells has been rapidly narrowing. Table 6.1, shows the comparison between these two technologies on the basis of their cell capacity, efficiency, and so on [130].

6.4 List of Solar Cell Materials

The following is a list of possible solar cell materials used to fabricate PV devices.

- Silicon is an easily available element in the earth's crust (another one is carbon [such as nanotubes, diamond-like-carbon (DLC)]).
- Binary compounds such as Cu_2S, Cu_2O, Cu-C, CdTe, CdSe, GaP, GaAs, InP, ZnP, a-Si: H, TiO_2, and so on.
- Ternary compounds such as Cu-In-S, Cu-In-Se, Cu-Zn-S, CdZnSe, CdMnTe, Bi-Sb-S, Cu-Bi-S, Cu-Al-Te, Cu-Ga-Se, Ag-In-S, Pb-Ca-S, Ag-Ga-S, Ga-In-P, Ga-In-Sb, and so on.
- Organic materials such as organic polymers, dyes, and so on.

6.5 Summary and Present Scenario of Crystalline Silicon Cells

1. It is well known that the reported efficiency of c-Si solar cells has reached ~25%; hence, the observed results show a better understanding of the junction properties and further innovations in cell design and fabrication technologies, and so on.

2. The efficiency gap between the best cell modules and the production of a module varies with advancement in technology; it is 10% lower

at every step, and therefore, manufactured solar cells have as little as 50% of the efficiency compared with the best laboratory cells.

3. As per the published data in the literature, PV production corresponded to 7900 MW all over the world in 2009. This was achieved using only single-crystal and polycrystalline silicon technology.

4. A decrease in the cost of a solar cell, from $100 to about $3/Wp (watt peak capacity), was observed due to increasing production of Si-PV from 200 kW in 1976 to 6900 MW in 2008. In the global PV market, the estimated annual growth rate of PV installations was approximately 40% from 2010 to 2016.

5. Therefore, considering the present fabrication technology and the cost of materials, it is seen that the cost of Si cells cannot be reduced significantly unless major innovations in the production of appropriate-quality silicon thin sheets take place.

6. At present, the technology uses 8″ or larger pseudo squares of ~200 μm thickness, with an efficiency of ~15%–16%. The energy is expected to be 16-5 kWh/Wp, the payback time of such cells is ~3–4 years, and the lifetime of the module is about 25 years.

7. Today, specially designed silicon solar cells having a cell efficiency of about 18%–20% are being manufactured in industries on a limited scale for special applications.

8. However, polycrystalline silicon solar cells with cell efficiency in the range of 12%–14% are being produced on large scale.

9. Today, specially designed thin film (approximately ~20 μm) silicon solar cells with an efficiency of about ~12% have been fabricated. Also, the production of hybrid thin film Si cells on the megawatt scale is being pursued.

10. Silicon solar cells currently dominate the PV market due to their fast-growing PV power, the production of more modern power panels are essential to match the demand in the energy sector.

11. Today, China and Taiwan are leading in the production of PV modules with a share of 68%, as compared with the other Asian countries, which have a share up to 14%, whereas Europe and the United States are contributing 4% and 6%, respectively.

References

1. R. Niishiro, H. Kato, A. Kudo, *Phys. Chem. Chem. Phys.* 7 (2005) 2241.
2. R. Eisenberg, D. G. Nocera, *Inorg. Chem.* 44 (2005) 6799.
3. E. Cuce, S. B. Riffat, C. H. Young, *Energy Convers. Manag.* 96 (2015) 31.

4. E. Cuce, C. H. Young, S. B. Riffat, *Energy Build* 86 (2015) 595.
5. E. Cuce, C. H. Young, S. B. Riffat, *Energy Convers. Manag.* 88 (2014) 834.
6. B. O'Regan, M. Grätzel, *Nature* 353 (1991) 737.
7. J. N. De Freitas, V. C. Nogueira, B. I. Ito, M. A. Soto-Oviedo, C. Longo, M. A. De Paoli, *Int. J. Photoenergy* 2006 (2006) 1.
8. J. Qiu, M. Guo, X. Wang, *ACS Appl. Mater. Interfaces* 3 (2011) 2358.
9. A. Hagfeldt, G. Boschloo, L. Sun, L. Kloo, H. Pettersson, *Chem. Rev.* 110 (2010) 6595.
10. M. K. Nazeeruddin, E. Baranoff, M. Grätzel, *Sol. Energy* 85 (2011) 1172.
11. A. Yella, H. W. Lee, H. N. Tsao, C. Yi, A. K. Chandiran, M. K. Nazeeruddin, *Science* 334 (2011) 629.
12. J. Lim, M. Lee, S. K. Balasingam, J. Kim, D. Kim, Y. Jun., *RSC Adv.* 3 (2013) 4801.
13. S. K. Balasingam, M. G. Kang, Y. Jun, *Chem. Commun.* 49 (2013) 11457.
14. S. K. Balasingam, M. Lee, M. G. Kang, Y. Jun, *Chem. Commun. (Camb.)* 49 (2013) 1471.
15. J. Wang, X. Zhang, Q. Su, *Rare Earth Solar Spectral Convertor for Si Solar Cells*, 2(139) (ISBN 978-981-10-1589-2) Springer Science + Business Media Singapore Pte Ltd. (2016).
16. B. S. Richards, *Sol. Energ. Mat. Sol. C* 90 (2006) 2329.
17. M. Grätzel, Photovoltaic and photoelectrochemical conversion of solar energy. *Philos. Trans. A Math. Phys. Eng. Sci.* 365 (2007) 993.
18. T. Trupke, M. A. Green, P. Würfel, *J. Appl. Phys.* 92 (2002) 1668.
19. W. B. Hung, T. M. Chen, *Sol. Energy Mater. Sol. Cells* 133 (2015) 39.
20. X. Chen, L. Liu, F. Huang, Soc. Rev., 44 (2015) 1861.
21. J. Liu, K. Wang, W. Zheng, W. Huang, C. H. Li, X. Z. You, *Prog. Photovolt. Res. Appl.* 21 (2013) 668.
22. W. Jing, Z. Xuejie, S. Qiang, *Rare Earth Solar Spectral Convertor for Si Solar Cells.* 139 (ISBN 978-981-10-1589-2), Up Conversion NanoParticles, Quantum Dots and Their Applications, Ru-Shi Liu, Editor, Volume 2 Springer (2016).
23. R. A. Kerr, Science 318 (2007) 1230.
24. N. S. Lewis MRS Bull. 32 (2007) 808.
25. G. W. Crabtree and N. S. Lewis, Phys. Today, 60 (2007) 37.
26. B. O'Regan and M. Grätzel, Nature, 353 (1991) 737.
27. T. W. Hamann, R. A. Jensen, A. B. F. Martinson, H. Van Ryswyk, and J. T. Hupp, Energy Environ. Sci., 1 (2008) 66.
28. F. Hao, C. C. Stoumpos, D. H. Cao, R. P. H. Chang, and M. G. Kanatzidis, Nat. Photonics, 8 (2014) 489.
29. S. Mathew, A. Yella, P. Gao, R. Humphry-Baker, C. F. E, N. Ashari-Astani, I. Tavernelli, et al., Nat. Chem., 6 (2014) 242.
30. M. A. Green, A. Ho-Baillie and H. J. Snaith, Nat. Photonics, 8 (2014) 506.
31. W. S. Yang, B. W. Park, E. H. Jung, N. J. Jeon, Y. C. Kim, D. U. Lee, et al., Science 356 (2017) 1376.
32. G. Grancini, C. Roldán Carmona, I. Zimmermann, E. Mosconi, X. Lee, D. Martineau, et al, Interface Engineering 8 (2017) 15684.
33. Y. Yang and J. You, Nature 544 (2017) 155.
34. Q. Chen, H. Zhou, T.-B. Song, S. Luo, Z. Hong, H.-S. Duan, et al. Nano Lett. 14 (2014) 4158.
35. B. R. Sutherland, S. Hoogland, M. M. Adachi, C.T.O. Wong, E. H. Sargent, ACS Nano, 8 (2014) 10947.

36. G. Nasti, S. Sanchez, I. Gunkel, S. Balog, B. Roose, B. D. Wilts, et al., Soft Matter, 13 (2017) 1654.

37. L. D. Pulfrey, *Photovoltaic Power Generation*. New York, NY: Van Nostrand Reinhold Co. ISBN 9780442266400 (1978).

38. N. Jenny, *Mater. Today* 14 (2011) 462.

39. J. Luther, M. Nast, M. N. Fisch, D. Christoffers, F. Pfisterer, D. Meissner, J. Nitsch, *Solar Technology*, Weinheim: Wiley-VCH (2008).

40. M. Jørgensen, K. Norrman, F. C. Krebs, *Sol. Energy Mater. Sol. Cells* 92 (2008) 686.

41. P. Riccardo, C. Chiara, B. Andrea, T. Francesca, C. Nadia. *Sol. Energy Mater. Sol. Cells* 100 (2012) 97.

42. M. C. Scharber, D. Mühlbacher, M. Koppe, P. Denk, C. Waldauf, A. J. Heeger, C. J. Brabec, *Adv. Mater.* 18(6) (2006) 789.

43. P. N. Rivers, *Leading Edge Research in Solar Energy*. New York, NY: Nova Science Publishers. ISBN 1600213367 (2007).

44. P. A. Troshin, H. Hoppe, J. Renz, M. Egginger, J. Y. Mayorova, A. E. Goryachev, A. S. Peregudov, et al., *Adv. Funct. Mater.* 19 (2009) 779.

45. T. Ameri, G. Dennler, C. Lungenschmied, C. J. Brabec, *Energy Environ. Sci.* 363 (2009) 347.

46. D. Duche, E. Drouard, J. Simon, L. Escoubas, P. Torchio, J. L. Rouzo, S. Vedraine, *Sol. Energy Mater. Sol. Cells* 95 (2011) 18.

47. C. Min, J. Li, G. Veronis J. Lee, S. Fan, P. Peumans, *Appl. Phys. Lett.* 93 (2010) 133302.

48. S. J. Tsai, M. Ballarotto, D. B. Romero, W. N. Herman, H. C. Kan, R. J. Phaneuf, *Opt. Express* 18 (2010) A528.

49. A. P. Kulkarni, K. M. Noone, K. Munechika, S. R. Guyer, D. S. Ginger, *Nano Lett.* 10 (2010) 1501.

50. H. Shen, P. Bienstman, B. Maes, *J. Appl. Phys.* 106 (2009) 073109.

51. H. Hoppe, N. S. Sariciftci, D. Meissner, *Mol. Cryst. Liq. Cryst. (Phila. Pa.)* 119 (2002) 113.

52. H. A. Atwater, A. Polman, *Nat. Mater.* 9 (2010) 205.

53. H. Sai, H. Fujiwara, M. Kondo, *Sol. Energy Mater. Sol. Cells* 93 (2009) 1087.

54. V. E. Ferry, L. A. Sweatlock, D. Pacifici, H. A. Atwater, *Nano Lett.* 8 (12) (2008) 4391.

55. R. A. Pala, J. White, E. Barnard, J. Liu, M. L. Brongersma, *Adv. Mater. (Deerfield Beach Fla.)* 21 (34) (2009) 3504.

56. N. C. Panoiu, R. M. Osgood, Jr., *Opt. Lett.* 32 (19) (2007) 2825.

57. V. E. Ferry, M. A. Verschuuren, H. B. T. Li, E. Verhagen, R. J. Walters, R. E. I. Schropp, H. A. Atwater, A. Polman, *Opt. Express* 18 (2010) A237.

58. V. E. Ferry, M. A. Verschuuren, H. B. T. Li, R. E. I. Schropp, H. A. Atwater, A. Polman, *Appl. Phys. Lett.* 95 (18) (2009) 183503.

59. W. Bai, Q. Gan, F. Bartoli, J. Zhang, L. Cai, Y. Huang, G. Song, *Opt. Lett.* 34 (2009) 3725.

60. S. Mokkapati, F. J. Beck, A. Polman, K. R. Catchpole, *Appl. Phys. Lett.* 95 (5) (2009) 053115.

61. M. A. Green, A. Ho-Baillie, H. J. Snaith, The emergence of perovskite solar cells, *Nat. Photon.* 8 (2014) 506.

62. S. D. Stranks, H. J. Snaith, Metal-halide perovskites for photovoltaic and light emitting devices, *Nat. Nanotechnol.* 10 (2015) 391.

63. M. K. Nazeeruddin, H. Snaith, Methylammonium lead triiodide perovskite solar cells: A new paradigm in photovoltaics, *MRS Bull.* 40 (2015) 641.

64. U. Bach, Perovskite solar cells: Brighter pieces of the puzzle, *Nat. Chem.* 7 (2015) 616.

65. M. Sessolo, H. J. Bolink, Perovskite photovoltaics: Hovering solar cells, *Nat. Mater.* 14 (2015) 964.

66. L. Wang, W. Fu, Z. Gu, C. Fan, X. Yang, H. Li, H. Chen, *J. Mater. Chem. C* 2 (2014) 9087.

67. J. H. Heo, H. J. Han, D. Kim, T. K. Ahn, S. H. Im, *Energy Environ. Sci.* 8 (2015) 1602.

68. J. P. Correa Baena, L. Steier, W. Tress, M. Saliba, S. Neutzner, T. Matsui, F. Giordano, et al., *Energy Environ. Sci.* 8 (2015) 2928.

69. Y. H. Deng, E. Peng, Y. C. Shao, Z. G. Xiao, Q. F. Dong, J. S. Huang, *Energy Environ. Sci.* 8 (2015) 1544.

70. D. Wang, M. Wright, N. K. Elumalai, A. Uddin, *Sol. Energy Mater. Sol. Cells* 147 (2016) 255.

71. Z. Gu, L. Zuo, T. T. Larsen-Olsen, T. Ye, G. Wu, F. C. Krebs, H. Chen, *J. Mater. Chem. A* 3 (2015) 24254.

72. T. M. Schmidt, T. T. Larsen-Olsen, J. E. Carlé, D. Angmo, F. C. Krebs, *Adv. Energy Mater.* (2015) http://dx.doi.org/10.1002/aenm.201500569.

73. W. S. Yang, J. H. Noh, N. J. Jeon, Y. C. Kim, S. Ryu, J. Seo, S. I. Seok, *Science* 348 (2015) 1234.

74. D. Bi, W. Tress, M. I. Dar, P. Gao, J. Luo, C. Renevier, K. Schenk, et al., *Sci. Adv.* 2 (2016) 1501170.

75. J.-H. Im, H.-S. Kim, N.-G. Park, *APL Mater.* 2 (2014) 081510.

76. L. Zuo, Z. Gu, T. Ye, W. Fu, G. Wu, H. Li, H. Chen, *J. Am. Chem. Soc.* 137 (2015) 2674.

77. Z. G. Xiao, C. Bi, Y. C. Shao, Q. F. Dong, Q. Wang, Y. B. Yuan, C. G. Wang, Y. L. Gao, J. S. Huang, *Energy Environ. Sci.* 7 (2014) 2619.

78. Z. Gu, F. Chen, X. Zhang, Y. Liu, C. Fan, G. Wu, H. Li, H. Chen, *Sol. Energy Mater. Sol. Cells* 140 (2015) 396.

79. Y. Li, *Sci. China-Chem.* 58 (2015) 830.

80. H. L. Hsu, C. P. Chen, J. Y. Chang, Y. Y. Yu, Y. K. Shen, *Nanoscale* 6 (2014) 10281.

81. G. E. Eperon, V. M. Burlakov, P. Docampo, A. Goriely, H. J. Snaith, *Adv. Funct. Mater.* 24 (2014) 151.

82. L. Huang, Z. Hu, J. Xu, K. Zhang, J. Zhang, Y. Zhu, *Sol. Energy Mater. Sol. Cells* 141 (2015) 377.

83. B. Tripathi, P. Bhatt, P. Chandra Kanth, P. Yadav, B. Desai, M. Kumar Pandey, M. Kumar, *Sol. Energy Mater. Sol. Cells* 132 (2015) 615.

84. B. Conings, L. Baeten, C. De Dobbelaere, J. D'Haen, J. Manca, H.-G. Boyen, *Adv. Mater.* 26 (2014) 2041.

85. B. Yang, O. Dyck, J. Poplawsky, J. Keum, A. Puretzky, S. Das, I. Ivanov, et al., *J. Am. Chem. Soc.* 137 (2015) 9210.

86. L. Wang, C. McCleese, A. Kovalsky, Y. Zhao, and C. Burda, J. Am. Chem. Soc. 136 (2014) 12205.

87. H. Zhou, Q. Chen, G. Li, S. Luo, T.-B Song, H.-S. Duan, Z. Hong, J. You, Y. Liu, Y. Yang, *Science* 345 (2014) 542.

88. H.-B. Kim, H. Choi, J. Jeong, S. Kim, B. Walker, S. Song, J. Y. Kim, *Nanoscale* 6 (2014) 6679.

89. C. Sun, Y. Guo, H. Duan, Y. Chen, Y. Guo, H. Li, H. Liu, *Sol. Energy Mater. Sol. Cells* 143 (2015) 360.

90. P. W. Liang, C. Y. Liao, C. C. Chueh, F. Zuo, S. T. Williams, X. K. Xin, J. Lin, A. K. Jen, *Adv. Mater.* 26 (2014) 3748.

91. C. Zuo, L. Ding, *Nanoscale* 6 (2014) 9935.

92. C.-C. Chen, Z. Hong, G. Li, Q. Chen, H. Zhou, Y. Yang, *Photon Energy* 5 (2015) 057405.

93. C.-C. Chueh, C.-Y. Liao, F. Zuo, S. T. Williams, P.-W. Liang, A. K. Y. Jen, *J. Mater. Chem. A* 3 (2015) 9058.

94. H.-L. Hsu, C.-C. Chang, C.-P. Chen, B.-H. Jiang, R.-J. Jeng, C.-H. Cheng, *J. Mater. Chem. A* 3 (2015) 9271.

95. S. T. Williams, F. Zuo, C. C. Chueh, C. Y. Liao, P. W. Liang, A. K. Jen, *ACS Nano* 8 (2014) 10640.

96. H. Yu, F. Wang, F. Xie, W. Li, J. Chen, N. Zhao, *Adv. Funct. Mater.* 24 (2014) 7102.

97. W. Zhang, M. Saliba, D. T. Moore, S. K. Pathak, M. T. Horantner, T. Stergiopoulos, S. D. Stranks, et al., *Nat. Commun.* 6 (2015) 6142.

98. J. Qing, H. T. Chandran, Y. H. Cheng, X. K. Liu, H. W. Li, S. W. Tsang, M. F. Lo, C. S. Lee, *ACS Appl. Mater. Interfaces* 7 (2015) 23110.

99. W. Qiu, T. Merckx, M. Jaysankar, C. MassedelaHuerta, L. Rakocevic, W. Zhang, U. W. Paetzold, et al., *Energy Environ. Sci.* 9 (2016) 484.

100. S. H. Chang, K.-F. Lin, H.-M. Cheng, C.-C. Chen, W.-T. Wu, W.-N. Chen, P.-J. Wu, S.-H. Chen, C.-G. Wu, *Sol. Energy Mater. Sol. Cells* 3 (2016) 375.

101. C. Huang, N. Fu, F. Liu, L. Jiang, X. Hao, H. Huang, *Sol. Energy Mater. Sol. Cells* 3 (2016) 231.

102. M. Liu, M. B. Johnston, H. J. Snaith, *Nature* 501 (2013) 395.

103. N. J. Jeon, J. H. Noh, Y. C. Kim, W. S. Yang, S. Ryu, S. I. Seok, *Nat. Mater.* 13 (2014) 897.

104. J. Burschka, N. Pellet, S.-J. Moon, R. Humphry-Baker, P. Gao, M. K. Nazeeruddin, M. Grätzel, *Nature* 499 (2013) 316.

105. Q. Chen, H. Zhou, Z. Hong, S. Luo, H. S. Duan, S. H. Wang, Y. S. Liu, G. Li, Y. Yang, *J. Am. Chem. Soc.* 136 (2014) 622.

106. M. Xiao, F. Huang, W. Huang, Y. Dkhissi, Y. Zhu, J. Etheridge, A. Gray-Weale, U. Bach, Y. B. Cheng, L. Spiccia, *Angew. Chem. Int. Ed.* 53 (2014) 9898.

107. A. Dualeh, N. Tétreault, T. Moehl, P. Gao, M. K. Nazeeruddin, M. Grätzel, *Adv. Funct. Mater.* 24 (2014) 3250.

108. J. Burschka, N. Pellet, S. J. Moon, R. Humphry-Baker, P. Gao, M. K. Nazeeruddin, M. Grätzel, *Nature* 499 (2013) 316.

109. G. Seo, J. Seo, S. Ryu, W. Yin, T. K. Ahn, S. I. Seok, *J. Phys. Chem. Lett.* 5 (2014) 2015.

110. W. Geng, L. Zhang, Y. N. Zhang, W. M. Lau, L. M. Liu, *J. Phys. Chem. C* 118 (2014)19565.

111. A. Muntasar, D. L. Roux, G. Denes, *JRNC* 190 (1995) 431.

112. A. Kojima, K. Teshima, Y. Shirai, T. Miyasaka, *J. Am. Chem. Soc.* 131 (2009) 6050.

113. Z. Xiao, C. Bi, Y. Shao, Q. Dong, Q. Wang, Y. Yuan, C. Wang, Y. Gao, J. Huang, *Energy Environ. Sci.* 7 (2014) 2619.

114. N. J. Jeon, H. G. Lee, Y. C. Kim, J. Seo, J. H. Noh, J. Lee, S. I. Seok, *J. Am. Chem. Soc.* 136 (2014) 7837.

115. S. Ryu, J. H. Noh, N. J. Jeon, Y. Chan Kim, W. S. Yang, J. Seo, S. I. Seok, *Energy Environ. Sci.* 7 (2014) 2614.

116. W. Fua, J. Yan, Z. Zhang, T. Ye, Y. Liu, J. Wu, J. Yao, C. Z. Li, H. Li, H. Chen, *Sol. Energy Mater. Sol. Cells* 155 (2016) 331.

117. M. A. Green, *Nat. Photon.* 8 (2015) 506.

118. J. Liu, R. Zhang, G. Sauve, T. Kowalewski, R.D. McCullough, *J Am Chem Soc* 130 (2008) 13167.

119. Z. Xiao, C. Bi, Y. Shao, Q. Dong, Q. Wang, Y. Yuan, C. Wang, Y. Gao, J. Huang, *Energy Environ. Sci.* 7 (2014) 2619.

120. N. J. Jeon, H. G. Lee, Y. C. Kim, J. Seo, J. H. Noh, J. Lee, S. I. Seok, *J. Am. Chem. Soc.* 136 (2014) 7837.

121. W. Fua, J. Yan, Z. Zhang, T. Ye, Y. Liu, J. Wu, J. Yao, C. Z. Li, H. Li, H. Chen, *Sol. Energy Mater. Sol. Cells* 155 (2016) 331.

122. W. S. Yang, J. H. Noh, N. J. Jeon, Y. C. Kim, S. Ryu, J. Seo, S. I. Seok, *Science* 348 (2015) 1234.

123. Z. Ren, A. Ng, Q. Shen, H. C. Gokkaya, J. Wang, L. Yang, W.-K. Yiu, et al., Sci. Rep. 4 (2014) 6752.

124. D. Bonnet, P. Meyers, *J. Mater. Sci.* 13 (1998) 2740.

125. K. Durose, P. R. Edwards, D. P. Halliday, *J. Cryst. Growth.* 197 (1999) 733.

126. D. Bonnet, *Thin Solid Films* 361 (2000) 547.

127. B. L. Williams, J. D. Major, L. Bowen, L. Phillips, G. Zoppi, I. Forbes, *Sol. Energy Mater. Sol. Cell* 124 (2014) 31.

128. L. Wang, C. McCleese, A. Kovalsky, Y. Zhao, C. Burda, J. Am. Chem. Soc. 136 (2014) 12205.

129. B. G. Mendis, L. Bowen, Q. Z. Jiang, *Appl. Phys. Lett.* 97 (2010) 092112.

130. www.civicsolar.com/support/installer/articles/thin-film-vs-crystalline-silicon-pv-modules

7

Phosphors for Environmentally Friendly Technology

7.1 Introduction

Green technology, also called *environmentally friendly technology*, is defined as technology that is friendly to our planet's environment. In this technology, special attention has been paid to the development and production of clean energy sources, which are not harmful to our environment, unlike the traditionally used fossil fuel energy sources. With the help of this green concept, we can conserve traditional fuel resources. It helps to reduce the level of harmful pollutants in the environment, which show adverse effects on human life and some other living species. Today, it is seen that specially in developing countries, the demand for basic needs is increasing due to the increasing population, and also, tremendous growth is observed in industry due to the high demand for new low-cost products that are manufactured using waste materials such as electronic and plastic raw materials followed by the process of separating and remanufacturing to make new products. Recycling is also helpful for waste management and provides useful resources. It is a key component of reducing the level of all types of waste in modern life. Therefore, to fulfill the demand for low-cost equipment produced from waste materials, the numbers of recycling plants are increasing. Along with this, the demand for electricity for regular household purposes and in the industrial sector is also increasing. Due to this, tremendous growth is observed in power generation plants that are based on coal and oil-based raw materials. However, they require a thousand tons of raw materials yearly to generate energy, thereby enhancing the level of toxic gases in the environment, which indicates their negative impact on our climate. Thus, considering this challenge in the world of globalization, there is an urgent demand to accept new and renewable sources for energy generation for the sustainable development of the nations. Green technology has more advantages in the fields of environmental science and technology and green chemistry and helps in the conservation of energy by using modified electronic devices and equipment. Indirectly, we can say that it is the core part of sustainable development.

There are only a few technologies that are playing an important role in balancing the environment. These include process recycling, wastewater treatment to remove harmful chemicals and dyes from water, absorbing air impurities by air purification processes, the use of naturally available sources for energy applications, and so on. In this textbook, we have mainly focused on the advantages of phosphors for the development of sustainable green energy and technology, which will have a positive impact on our health and environment. One approach that falls into this category is the use of eco-friendly non-toxic phosphor materials for the development of solid-state lighting technology, regarded as an energy-efficient source for the 21st century, and another is the use of upconversion and downconversion phosphor layers to improve the current conversion efficiency of Si-based photovoltaic (PV) technology. Today, many research laboratories worldwide are working on the replacement of traditional energy sources such as fossil fuel technology by using renewable energy sources (natural resources) such as sunlight, wind, geothermal, nuclear, biodiesel, biogas, and so on. Thus, seeing the effect of globalization on climate change, it is our responsibility to make the public aware of this and to adopt the concept of "Go Green" by using friendly technology. Green technology uses a diverse range of science and technology to provide the knowledge and connection between the use of human technology and its effect on the climate and natural resources. This is the broad area which can protect the environment. Also, it has an eco-friendly and sustainable approach that relates to the development of the economy of the country. The main goal of this technology involves identifying and solving nature-related problems. Also, it has long-term advantages. It will stabilize the balance between energy conservation and the use of resources. The main objectives of green technology are as follows:

- To find approaches to sustainable development.
- To create products that can be reused as new.
- To reduce the level of energy consumption and its component of global warming.
- To find alternatives to existing technologies.
- To develop economic activity around technologies that are beneficial to protecting our lovely planet.

So, it is an energy-efficient technology that gives more benefits with fewer inputs, according to the Kyoto Protocol (1997) regarding the UN framework convention on climate change, in which the focus is on some big issues, such as how to reduce the level of carbon emissions, which are the main factor in global warming, which is rapidly changing our climate; how to stop the rapid increase in deforestation all over the world; and how to promote the semi-governmental or nongovernmental organization of countries to work for a low-carbon society through green growth by adopting the concept of green energy resources and technology. Indirectly, we have to develop an eco-friendly technology that has a good impact on the environment. This is possible only when we first promote the use of renewable resources by considering the social aspects. In the near future, in the globalized world, the whole world is facing big problems in environmental safety regarding our daily needs, such as a clean environment and clean drinking water, and controlling our climate, which has the greatest effect on agriculture, which totally depends on the weather, as is seen in developing countries; thus, the basic challenge facing the field of agriculture is to produce enough food to fulfill the demand of a growing population. But, due to the imbalance in the climate, most agricultural food crops have been damaged yearly, and farmers cannot cover the annual demand for food. As per the current scenario and changes observed in the agricultural food production system, food safety and management is also a major issue in the 21st century. The only solution that has the capability to resolve all issues concerning the environment is green technology, a potential way of sustaining mankind and nature. The following subsections discuss the types of sustainable and renewable resources for the development of green energy and technology.

7.1.1 Types of Renewable Energy Sources

7.1.1.1 Solar Energy

Solar energy is the light energy emitted by the sun, which is used by different technologies, such as solar heaters, PV, solar thermal energy, solar architecture, and so on [1]. Hence, it is an evergreen source for the development of renewable energy and related technologies; these are broadly classified as passive or active, depending on how the technology is used to capture and distribute solar energy or convert it into solar power. Active solar technology is based on PV systems, in which it uses concentrated solar power and solar water heaters to harness solar energy. Passive technology is based on PV solar film, which may be spread over windows, walls, roofs, and floors of a building to capture the maximum solar radiation incident on it and convert it into energy. Recently, solar energy has become an ideal, most promising, clean and pollution-free source of energy [2]. It is called a *universal natural source* and an *infinite reserve*, and it has the further advantage of being a clean and economical source for sustainable development. For this reason, solar energy has become one of the most important sources of energy in the 21st century [3, 4]. In 2011, the International Energy Agency reported that "the development of affordable, inexhaustible and clean solar energy technologies will have huge longer-term benefits." So, the use of solar energy will fulfill the need for energy security, help to sustain the economy, reduce pollution, reduce global warming, and help to decrease the use of fossil fuels-based pollutants for energy. These are a few of the global advantages of solar energy. Another important study from the frontiers of science has found that the increase in air pollution from natural dust particles and other pollutants from industry, coal power plants, cars, heavy vehicles, and so on has slashed the power conversion efficiency of Si solar cell panels by 17%–25% across parts of underdeveloped countries in Asia, because the atmospheric blocks the sunlight from reaching solar panels. And if the particles are deposited in large quantities on a panel's surface, they minimize the conversion efficiency of the panel.

7.1.1.2 Wind Energy

Wind is defined as the perceptible natural movement of the air. It is caused by different physical phenomena such as uneven heating of the atmosphere by the sun's radiation, irregularities on the earth's surface, and the motion of the earth [5]. Therefore, wind energy is also a form of solar energy. The patterns of wind flow are modified by the earth's terrain, bodies of water, vegetative cover, and so on. Therefore, the motion of wind energy is helpful to run modern turbines at higher speeds; wind turbines convert the kinetic energy in the wind into mechanical power. Later, this mechanical power can be converted into electricity for household, school, and small-scale purposes. A wind turbine work in the opposite way to a fan; that is, instead of using electricity to generate a wind, it produces electricity by using a wind [6]. Today, energy consumption has been mostly dependent on fossil fuel sources for a long time, which creates many problems related to health, climate change, and environmental pollution. Thus, considering this issue, wind energy is a clean and environmentally friendly alternative source to replace fossil fuel–based energy sources. Therefore, it is important to develop an appropriate and advanced wind speed model to produce the maximum wind energy. Wind power is one of the most popular forms of renewable energy source. Despite the many advantages of wind technology, this kind of renewable energy source has faced more challenges for system operators, especially from the voltage stability point of view. Also, installation cost is another factor that should be considered with respect to its impact on the voltage stability. Hence, the economic problems of wind energy use should be considered in addition to its operational and technical impacts [7].

7.1.1.2.1 Advantages and Disadvantages of Wind Energy

7.1.1.2.1.1 Non-Polluting Resource Wind energy is a free, non-polluting, and friendly source of electricity, and to date, its use has been limited. Unlike different conventional energy sources such as fossil fuel–based power plants, it does not emit harmful pollutants and increase the level of greenhouse gases in the atmosphere. But it requires the right geographical conditions and airflow quality to generate electricity.

7.1.1.2.1.2 Cost Issues Recently, the cost of wind power has decreased dramatically from its cost over the past 10 years due to advancement in technology. But initially, it requires a higher investment than fossil-fuel generators.

Roughly 80% of the cost is used by machinery, and the remainder is used in site preparation and installation. If wind-generated energy is compared with a fossil fuel–based energy system on a cost basis, it is found that the wind generating system is more beneficial, because there is no need for fuel, and it requires minimal operating expenses.

7.1.1.2.1.3 Environmental Concerns Although wind power plants have a relatively low impact on the environment compared with fossil fuel power plants, there is some concern over the noise produced by the rotor blades, aesthetic (visual) impacts, and birds and bats having been killed (avian/bat mortality) by flying into the rotors. Most of these problems have been resolved or greatly reduced through technological development or by properly siting wind plants.

7.1.1.2.1.4 Supply and Transport Issues The major challenge of using wind as a source of power is that it is intermittent and does not always blow when electricity is needed. Wind cannot be stored, and not all winds can be harnessed to meet the timing of electricity demands. Further, good wind sites are often located in remote locations far from areas of electric power demand, which raises big transportation issues. Finally, wind resource development may compete with other uses for the land, and those alternative uses may be more highly valued than electricity generation. However, wind turbines can be located on land that is also used for grazing or even farming.

7.1.1.3 Geothermal Energy

The world's oldest geothermal system is found in Chaudes-Aigues, France, which has well-known hot spring waters that have been operating since the 14th century [8]. The generation of geothermal energy does not require fuel, except for pumps, so it controls fuel-based pollution. However, the power generation costs are significant [9]. The generation of energy using the earth's internal heat, which originates from the time of formation of the planet and is due to the radioactive decay of materials, is called *geothermal energy* [10, 11]. This energy has been used for many years in some countries because of its important applications in cooking and heating. This technology uses the thermal energy inside the rock and fluids beneath the earth's crust, which are situated several miles below the surface; further down, there is an extremely hot molten rock called *magma*. To produce electricity, a well is drilled into underground reservoirs to tap steam and very hot water, which moves turbines and generates electricity. The generation of electricity was done for the first time in Larderello, Italy, in 1904. There are now three types of geothermal power plants: dry steam, flash, and binary. Dry steam is the oldest geothermal technology, which takes steam from fractures in the ground and uses it directly to rotate the turbine. Flash plants pull deep, high-pressure hot water into cooler, low-pressure water.

The steam produced by this process is used to rotate the turbine. In binary plants, the hot water passes by a secondary fluid with a much lower boiling point than water. This causes the secondary fluid to turn to vapor, which then rotates the turbine. Most geothermal power plants in the future will be binary plants. Today, more than 20 countries are generating geothermal energy [12]. Currently, the United States is the world's largest producer of geothermal energy. In Iceland, swimming pools and many buildings are heated with the help of geothermal hot water. Iceland possesses 25 active volcanoes and many hot springs and geysers [12].

The following subsections list some advantages and disadvantages of geothermal energy compared with other forms of renewable energy resources.

7.1.1.3.1 Advantages

The advantages of geothermal energy are as follows:

- It can be produced without burning a fossil fuel (e.g., coal, gas, or oil).
- A geothermal plant produces only about one-sixth of the carbon dioxide that another relatively clean, natural fuel-based power plant would produce.
- Binary geothermal plants do not release carbon or harmful gas emissions.
- As compared with other renewable sources, such as solar and wind energy, it is always available throughout the year.

7.1.1.3.2 Disadvantages

- The plant design and construction are relatively expensive.
- There are some environmental problems, such as the release of hydrogen sulfide and the disposal of some geothermal fluids containing low levels of toxic materials.

Thus, based on these advantages and disadvantages of geothermal energy, it has the capability to provide heat for many decades; eventually, specific locations may cool down [12].

7.1.1.4 Hydroelectric Energy

In this case, electricity is generated from flowing water, so that it is called *hydroelectric energy*. This is also a form of renewable energy resource. It is a natural source of energy and a very important energy source at both national and international level. All over the world, approximately 19% of electricity is generated from hydroelectric energy, and among other countries, the United States has the potential to generate about 7% of the power. Currently, more than 25 countries in the world generate 90% of their electricity based on hydropower, and of these, 12 countries are 100% based on hydro. Thus, it is a clean, environmentally friendly, mature, reliable, and cost-effective source of power generation. The capacity of the power plant depends on the capacity and storage of the reservoir. Hydropower dams with large reservoir storage can be used to store energy over time to meet the requirements. The following are some key advantages and disadvantages of hydropower generation.

7.1.1.4.1 Advantages

- It is a powerful and globally important clean source of energy generation.
- It can meet load fluctuations, such as those observed in conventional thermal power plants, minute by minute.

7.1.1.4.2 Disadvantages

- A large financial investment is required to set up a power generation plant.
- A large water reservoir is needed, but due to present global warming issues in many developing countries (e.g., India), rainfall and storage of water in the reservoir are seriously affected, which is a major drawback.

7.1.1.5 Biomass Energy

This type of energy generation source has been used from ancient times, when people used to burn wood or coal to generate heat. These are the most common sources to generate biomass energy. Besides these, other products used to generate energy include crops, plants, landfill, municipal and industrial waste, trees and agricultural waste, and so on. Therefore, biomass is also a renewable source of energy. The advantages and disadvantages of the biomass energy generation are as follows.

7.1.1.5.1 Advantages

- It does not produce greenhouse gases, and also, it can be easily extracted through the combustion process.
- It is helpful for waste management and also helps to reduce the problem of landfill.

7.1.1.5.2 Disadvantages

- Biomass is a comparatively inefficient source as compared with fossil fuels and other sources of energy.
- It releases methane gas, which is harmful to the environment.

7.2 Advantages of Phosphors

7.2.1 Need for Energy-Saving Technology

Recently, the demand for electricity has been rising rapidly due to a vast increase in industry and population in developing countries. Lighting devices were the first service component offered by electric utilities for indoor and outdoor applications, and currently, they have more electrical end-uses. According to a worldwide global database, lighting consumes 650 Mt of energy and emits around 1900 Mt of CO_2 for this amount of energy generation through a power plant, which is higher than the 70% of CO_2 emissions coming from passenger vehicles all over the world and also three times higher than emissions from aircraft [13]. Thus, it is seen that the lighting sector consumes almost one-fifth of the total amount of electricity generated. Almost 43% of global lighting electricity is consumed by the commercial sector, while the residential sector consumes 31% in lighting, the industrial sector represents 18% of electricity consumption, and the rest of the outside sector is responsible for at least 8% of electricity consumption [13]. Thus, energy-efficient technology is essential for solving these problems. It may be beneficial to save energy and the environment by reducing the level of greenhouse gases. The European Union is committed to adopting a new energy policy and developing technology that will enhance the energy efficiency of the existing devices by up to 20% by 2020. The proposed measures require minimum energy–using energy-saving equipment in different areas such as buildings, industry, transport, and energy generation plants. The energy-saving potential of solid state-devices (light-emitting diodes [LEDs]) is much higher than for the lighting devices available in the market.

7.2.1.1 Solid-State Lighting (SSL) Technology

Lighting, artificial or natural, plays a most important role in human life for carrying out our day-to-day living activities. Every task of our life requires an appropriate lighting level, which to a large extent depends on our needs and the available lighting level. In general, the level of illumination recommended for household purposes ranges from 50 to 150 lux. However, it is not possible for everyone to access lighting sources with illumination within this range. In developing countries, it is not possible to use an artificial lighting due to the growing population, so people residing in remote areas use oil-based lamps for household purposes, because lamps that provide better illumination are too expensive for them to afford and maintain. Today, there is a challenge to develop highly efficient, reliable, and affordable lighting devices that have the potential to replace oil lamps with environmentally friendly lighting sources. The journey of lighting started with oil lamps and continued through incandescent and then fluorescent lamps, which are today being replaced by solid-state LEDs, which have energy-saving potential. Also, oil-based lamps are a major source of greenhouse gas, and they consume a significant amount of energy, as well as causing fire accidents and so on. According to a global survey, this type of lamp uses about 470 million barrels of oil per year and damages the environment by releasing approximately 400 billion pounds of CO_2 gas into the earth's atmosphere annually. So, there is a direct impact on climate change, and this factor is responsible for global warming. To avoid such losses and to develop national economies, there is an urgent requirement to accept an environmentally friendly technology. SSLs, as an emerging technology, have the potential to address all these challenges. The term SSL refers to a type of lighting technology that includes semiconductor LEDs, organic LEDs (OLEDs), and polymer light-emitting diodes (PLEDs) as sources of illumination rather than electrical filaments and plasma, which is used in arc lamps such as fluorescent lamps, and so on. If we look at the historical background of lighting devices, we first focus on fluorescent lamps, which delivered 64% of light, having efficacy in the range 40–100 lm/W. They are mostly used in general-purpose applications such as indoor lighting and industrial buildings. So, these lamps accounted for 20% of global sales and consumed 45% of electricity. Following this, high-intensity discharge (HID) lamps were introduced, such as mercury vapor lamps, metal halide lamps, high- and low-pressure sodium lamps, and so on. Such lamps give a large amount of light but have efficacy in the range 35–150 lm/W. They are mostly used in outdoor lighting, such as street lighting, and indoor lighting in some industries. They accounted for 1% of global lamp sales, and consumed 25% of electricity, and provided 29% of the output light. Mercury vapor lamps are an old and inefficient lighting technology that has a low cost as compared with the other alternative sources of lighting. Thus, compared to the use of High-intensity discharge (HID) lamps, the use of fluorescent lamps are more beneficial, due to their positive impact on the overall energy consumption of lighting. [14]. The word *solid-state* corresponds to light emitted by the process of solid-state electroluminescence as opposed to the mechanism observed in incandescent bulbs or fluorescent tubes. SSL has more advantages than incandescent lighting technology. It has the ability to generate visible light with reduced heat generation and lower energy consumption. Recently, this SSL technology has been used in many more areas, such as traffic lights, modern vehicle lights, street and parking lights, train marker lights, building exteriors, remote controls, and so on. It has also made significant advances in industry. Currently, SSL technology based on LEDs is a growing technology with greater energy-saving potential than traditionally used lamp-based lighting systems. LEDs were first commercialized in 1960s, and currently, they have replaced most lighting devices with SSL. This technology can save energy and money as compared with other existing devices such as incandescent and fluorescent lamps, and so on. Since 2013, SSL devices have been available in stores, and their cost is also somewhat lower, so that they are attracting many consumers. LED technology is well known due to its superior performance and advantages of energy efficiency, optical efficiency, lower weight, compact design, and good flexibility. Because of their good flexibility, LEDs can be allowed in many applications where standard lamps cannot be used [15]. SSL development requires further research and innovation to discover new kinds of materials with good physical and chemical properties, high quantum efficiencies (QEs), high stability, and long durability, with many applications in the areas of display science and technology. In the following section, we see the advantages and disadvantages associated with LED technology and a few general applications of SSL.

7.2.1.1.1 *Advantages of LEDs*

The advantages of LEDs as compared with other light sources are listed here [16]

- High brightness
- Low operating voltage required
- Small weight and compact nature
- Long lifespan
- Low heat emission
- Mercury free
- High luminous efficacy
- Flexible
- Available in different colors
- No ultraviolet or infrared radiation
- Cool light emission

7.2.1.1.2 *Disadvantages of LEDs*

The following are the few disadvantages of LEDs.

- Lack of standard test conditions
- Higher price
- Low power due to small lamp size
- Need for thermal management to enhance the lifetime of the device
- Blue pollution (blue light, which causes damage to the retina)
- Temperature dependence (with increase in device temperature, the performance changes) [16–19]

7.2.1.1.3 *Potential Advantages of SSL*

SSL is an emerging field of the 21st century and has captured most areas in display technology, such as mobile phones, TVs, indicator automotive displays, and so on. Currently, these lights are used for decorative and architectural lighting purposes because of their superior color and good spectral distribution in the visible region, and the devices have a low maintenance cost. As well as its energy-saving potential, SSL can offer

- Less radiated heat
- Production of lower waste heat
- Low maintenance costs
- Applicability for street lighting
- Cost-effectiveness

7.2.1.2 Environmental and Health Impacts of SSL

LEDs are an emerging environmentally friendly lighting source due to some important environmental advantages as compared with incandescent and fluorescent light sources. Another advantage of LEDs is that they require a low operating voltage, which indicates energy-saving capabilities and also helps to reduce the level of CO_2. They do not contain any toxic materials: glass, filaments, mercury, and so on. Considering these benefits, LEDs are a much safer alternative to the existing lighting technologies that are used in commercial and industrial applications [20–22]. Hence, they are an important component of economic and environmental sustainability; they are cost-effective technologies and help to reduce greenhouse gases in the environment. However, it is found that LEDs consist of semiconductor chips made up of arsenic, gallium, indium, and/or antimony coupled with light-emitting phosphor materials; such materials are toxic, with potential impacts on the environment and human health and ecological toxicity effects when disposed of at the end of their service time. They also contain some other metals, such as copper, gold, nickel, and aluminum, which are non-toxic. So, it is necessary to develop a proper way to recycle these materials (gallium, indium, and so on) and to develop new LED products considering this drawback. Another health risk related to the lighting technology is that it emits light in different spectral ranges, and white light can be obtained by mixing suitable color-emitting materials as compared with other ordinary light sources. However, this may cause some health problems; the light is potentially harmful to the eye due to the presence of a blue component in white light, and also, they exhibit some electromagnetic radiation in surrounding areas. Highly efficient lumen output LED bulbs are currently used in all lighting devices because of their interior or exterior applications in the daytime or at night. However, researchers are busy finding better visual performance benefits of lighting sources for interior and exterior use. Similarly, they have recommended practically proved specification values and restrictions applicable for minimum daytime and maximum night-time exposures to light [23]. Thus, there is a need to find potential substitutes for the existing devices by replacing the materials with new, energy-efficient materials that have a positive impact on climate change, have energy-saving potential, are non-toxic in nature, and show a low health risk during their operation.

7.2.2 Energy Conversion Technology

Recently, saving energy has become a priority, and it is the most important topic in the world of globalization. Conversion science and technology is the basic concept that deals with the renewable and sustainable development of energy resources. The energy conversion phenomenon plays an important role in solar cell devices; that is, the conversion of sunlight into electricity, or in the case of compact fluorescent lights, the conversion of incident UV light into visible light. For this application, rare earth–doped phosphors are excellent candidates for converting short-wavelength light into a longer wavelength. Also, a suitable coating of nanophosphor materials in energy conversion devices may help to enhance the device performance and to achieve high energy efficiency. The current attention to energy saving and reduction of CO_2 emissions in the atmosphere should, therefore, give an additional boost to the development of LEDs for lighting. The remaining disadvantages of LEDs are the need for extensive cooling of high-power devices (ultimately limiting the maximum power per LED chip), the need for current driving, and the lack of high–color quality white LEDs. Such white LEDs are typically made by starting with a blue-emitting LED and converting part of its light to green and red by means of one or more phosphor materials. There are only a limited number of phosphor materials known that are suitable for this wavelength conversion. The present review is focused on six main performance requirements for state-of-the-art color conversion phosphors:

- An emission spectrum that, in combination with the emission of the other components (LED, other phosphors), leads to a pure white emission with a specific color rendering and color temperature
- An excitation spectrum showing good overlap with the pumping LED and large absorption strength
- An emission spectrum, excitation spectrum, and QE that remain unchanged at elevated temperature
- A QE approaching unity, thus maximizing the overall electrical-to-optical conversion efficiency of the entire LED–phosphor package
- Excellent chemical and temperature stability
- The absence of emission saturation at high fluxes

7.2.2.1 Si Solar Cell Technology

Solar energy is the technology used to harness the sun's energy and make it usable. By 2011, the technology fulfilled less than one-tenth of 1% of global energy demand. Recently, many research groups have been active in the field of PV in universities and research institutions around the world. Researchers have focused on three areas:

- Making current–technology (solar cells) cheaper and/or more efficient to effectively compete with other energy sources
- Developing new technologies based on new solar cell architectural designs
- Developing new materials to serve as more efficient energy converters from light energy into electric current or light absorbers and charge carriers

Now that we face the challenge of global warming, the development of new and advanced green energy materials has been an important issue in materials research. There is a need to use natural sources for energy generation, so among the different fields of energy, PV technology is an emerging field due to the availability of sunlight everywhere in the world. The PV cell is one of the promising devices that can be used to generate sustainable energy. Many efforts have been made to explore new materials that are able to enhance the power conversion efficiency (η) of present solar cell devices. The efficiency of a PV cell depends on wavelength of incident light and how much light is converted into electricity, and the η–λ relationship is characterized by the spectral response of the cell.

In general, PV does not have the ability to use all incident light for energy generation cells; only a small portion of the solar spectrum, corresponding to some of the UV-visible or near infrared (NIR) range is used for electricity, and part of the solar light is wasted due to a spectral mismatch between the incident light and the solar cell plate. To avoid the spectral mismatch and to minimize other losses that affect the efficiency of the cell, some new ideas have been introduced to enhance the efficiency of existing solar cells by using an upconversion or downconversion phosphor as an active layer by coating it onto the solar plate; their properties have already been discussed in Chapter 3. Another concept reported for enhancing the efficiency is the use of a luminescent solar concentrator (LSC). An LSC is a device for concentrating radiation, solar radiation in particular, to produce electricity. It operates on the principle of collecting radiation over a large area, converting it by luminescence and directing the generated radiation into a relatively small output target. In the case of light guide–based solar concentrator systems, the incident light is coupled to a plate and then guided toward small photovoltaic cells on the basis of the phenomenon of total internal reflection (TIR). Such devices are attractive, inexpensive, and thin and can also be easily integrated into appliances.

The major challenges for designing an efficient LSC are

- It requires a good luminescent material.
- Apart from the basic requirements (good absorption and high QE), the luminescent material should have low reabsorption.
- Another requirement is that the luminescent material should be applied in such a way that it does not scatter.

The main drawbacks of luminescent materials such as organic dyes [24] and quantum dots [25], which are used in LSCs, is that they suffer from reabsorption due to the significant overlap between their absorption and emission spectra. Hence, the rare earth compounds are more promising, because they have large shifts between absorption and emission; however, the optical absorption between 4f levels is forbidden. Therefore, both the inorganic phosphors [26] and the organo-metallic rare earth complexes [24] developed to date can absorb only a limited part of the solar spectrum. Therefore, when considering the optimal phosphor properties, one has to realize that these will be a compromise, because small reabsorption of energy indicates a large spectral shift between absorption and emission spectra, which means that the absorption spectra cannot extend too far into longer wavelengths range. However, if a light guide is used, which is made from a common polymer such as polymethylmethacrylate (PMMA) or polycarbonate, these materials can demonstrate strong absorption [27] above the infrared range.

Thus, in the case of an inorganic phosphor, the converted sunlight that scatters toward the solar cell contributes more than half to the cell efficiency. Hence, further research is required to optimize the performance of this LSC configuration to enhance the light conversion efficiency of the solar cell.

7.2.2.2 Scope and Advantages of Crystalline Silicon (c-Si) Solar Cells

At present, the c-Si solar cell possesses the dominant position in the modern PV marketplace due to developments in technology. It benefits from a huge amount of industrial investment and also the availability of a large elemental abundance of Si everywhere [28]. Solar cells made from polysilicon and monocrystalline silicon are the ideal semiconductor materials at present [28]. The c-SI solar cell covers 90% of market share, and it comprises the largest proportion of the cost of PV power, so that new manufacturing process technologies for preparing highly efficient c-Si solar cell materials at low cost have attracted much attention at both national and international level [29]. Of the two semiconductor materials discussed, polysilicon materials take up more than 70% of the market share of c-Si. Solar cells made from high-quality monocrystalline silicon have been much favored by the market [30]. Currently, a research and innovation team led by Prof. Yongnian Dai [31], Kunming University of Science and Technology, have found a new metallurgical process for developing high-efficiency c-Si solar cell materials at low cost and have shown that it is a direct and effective way to reduce the cost of PV power sources. It will help to make PV power cheaper than or equal to the cost of traditional energy sources such as thermal power, and so on.

c-Si solar cells have some major advantages, such as

- A high efficiency rate, up to 14%–22%
- High stability of silicon, so that it has been established as the technology leader in solar cell technology

7.2.3 Photocatalysts for Wastewater Treatment

All over the world, the industrial sector is growing fast, and at the same time, some major issues that disturb the environment need to be overcome. Problems related to water and air pollutants have reached the top position and attracted most attention everywhere in the world [32, 33]. Among these problems, special attention must be given to resolving the issues associated with wastewater treatment [34], because wastewater from textile and other industries contains many harmful contaminants, but the traditional technology used to treat such wastewater cannot remove antibiotics [35, 36], dyes [37, 38], organic pesticides [8, 9], or polycyclic aromatic hydrocarbons (PAHs) [39, 40] effectively. In the last few years, different types of method have been proposed and employed to remove harmful contaminants from wastewater. These include ozone treatment, electrochemical processes, direct decomposition of water, and photocatalysis, reported by Arslan Alaton et al. (2007), Rivas et al. (2012), Holkar et al. (2016), and Orge et al. (2017), respectively [41]. Ozone treatment works by directly introducing ozone gas into wastewater. Here, the gas directly reacts with organic dyes or compounds and indirectly results in decomposition. Compounds that remain decomposed undergo ozone treatment. Fenton oxidation technology produces OH by decomposing H_2O_2 by Fe^{2+} in liquid medium with or without light or UV irradiation. But this type of oxidation technology is not applicable for a medium with pH > 4. So, the operational cost of this method is also high (Barbusinski, 2009; Pouran et al., 2015). Direct decomposition is a method of OH generation in which water molecules are decomposed by external energy sources such as microwaves, ultrasound, and electromagnetic radiation. The water molecules break into H^+, OH, and electrons by irradiation. If the water is discharged by applying a pulse voltage with the help of high-voltage electrodes of the order of 10 kV placed nearer to each other, then, localized heat is generated, which starts decomposing the water molecules. Due to overheating, a hot spot is produced when irradiation occurs under ultrasound or microwave. The last method used for wastewater treatment, photocatalysis, is an excellent source of OH generation technology that includes large–band gap metal oxides such as $SrTiO_3$, $BiTiO_3$, WO_3, $ZnWO_4$, ZnO, CuS/ZnS, Bi_2OTi_2O, ZnS, Ag_2CO_3, Bi_2WO_6, Nb_2O_5,

Fe_2O_3, TiO_2, and so on. In this method, semiconductor materials act as photocatalysts for the further process, which is activated by incident light radiation (Gaya and Abdullah, 2008). The semiconductor photocatalyst (SP) consists of a large energy gap separated by valence bands (VB) and conduction bands (CB). However, the practical applications of heterogeneous photocatalysts such as TiO_2 and ZnO [41–43] will be limited due to their weak visible light photocatalytic activity and wide band gap nature; they exhibit low visible light responsiveness. Currently, researchers are focusing on the study of conjugated polymer–based photocatalysts due to their excellent photocatalytic activity, lower cost, and wide range of sources [44]. Graphitic carbon nitride (g-C_3N_4) is a well-known conjugated polymer–based photocatalyst that has been widely used in photocatalytic water splitting, the degradation of organic pollutants, and the removal of CO_2 because of its suitable band structure and shows an excellent response in visible light [45–47]. It is also applicable in conjugated polymer engineering materials and is widely used in the areas of separation membranes, microelectronics, nanomaterials, and so on [48]. Besides this, a new inorganic compound semiconductor has been proposed for wastewater treatment due to superior reports on its photocatalytic activities such as $BaMoO_4$ but it rarely found, due to its large band gap (4.2 eV) with rapid recombination of charge carriers (electrons and holes) [49]. The wide band gap indicates that it is active only under the UV light spectral region and shows poor photocatalytic activity; hence, it is further modified by designing plasmonic photocatalysts composed of silver or gold metals and semiconductors, so that this material exhibits plasmonic activity and shows strong visible light absorption due to its localized surface plasmon resonance [50, 51]. Therefore, because of the internal electric field, a Schottky junction is formed near the metal–semiconductor interface, which enables the recombination of photogenerated charge carriers in opposite directions to support the charge-separation efficiency to enhance the photocatalytic activity [52–54]. Along with this, a $BaMoO_4$ host doped with rare earth ions (rare earth = Er^{3+}/Yb^3) enhances the photocatalytic activity due to the increase in light absorption and increase of surface barrier resulting from doping with rare earth ions [55]. When this $BaMoO_4$:Er^{3+}/Yb^{3+} material is effectively excited by a 980 nm NIR wavelength, the Yb^{3+} ions absorb a large amount energy and transfer it to Er^{3+} ions, which show the NIR to visible upconversion properties. Therefore, Er^{3+} ions transfer the energy to the host material, which helps the dye degradation process [56–58]. Recently, new ideas have been proposed to improve the photocatalytic features of semiconductor materials by coupling them with light-emitting phosphorescent particles (Vaiano et al., 2014). Phosphors are solid luminescent materials consisting of optically active centers that can absorb incident light and emit a light of longer wavelength range (Li and Sakka, 2015). When a rare earth emitting phosphor is coupled with a suitable photocatalyst, they can absorb light coming from an external source, emit a specific wavelength of radiation in the core of the photoreactor, and transfer the additional photons to the photocatalyst. So that rare earth ions are useful to make photo catalyst active in the visible spectral region and improve the performance TiO_2 and related materials used for dye degradation [59].

7.3 Clean Technology: An Economics of Sustainable Development

The term *technology* implies the application of knowledge for practical purposes. The field of *green technology* encompasses a continuously evolving group of methods and materials, from techniques for generating energy to non-toxic cleaning products. This field has gained a lot of importance in the current century due to the new innovations and changes observed in daily life of a similar magnitude to the "information technology" explosion over the last two decades. Therefore, worldwide, few development issues have been overcome regarding the degradation of the local and global environment. If economics and environmentalism work together, this may help to enhance sustainable development in some developing countries. It is well known that with increased efforts in the direction of globalization, the level of competition among companies in various sectors of work also increases. This globalization in different fields provides new developments, and it has also increased the pace of development in many developing countries, such as India and China. Through globalization and development, we are continuously damaging our environment by producing greenhouse gases, water pollution, soil pollution, and so on. In this era of globalization, many scientists and researchers concerned with environment and ecology suggest that if this rate of exploitation continues, then the day is not so far off when our lovely planet and its environment will be not suitable for sustainable life. This is why the concept of green technology has been introduced, which uses technology in such a manner that on the one hand, development continues, and on the other hand, the level of negative environmental impact is reduced to a minimum. In this situation, many researchers feel that there is urgency in the direction of environmental and ecological stability. This is possible only by adopting green technology—technology that has the potential to improve environmental performance and stability relative to other technology. As discussed earlier, many countries worldwide are making efforts to conserve energy. This is possible only by using equipment or machinery that requires a lower amount of energy, resulting in low consumption of electricity and thereby reducing the use of fossil fuels to generate electricity. Both energy conservation and efficiency are energy reduction techniques.

7.3.1 Advances in Urban Life

Today, the term *green technology* is expanding very rapidly. Many countries are taking remarkable initiatives toward green technology and working together to resolve the issue of global warming. Many awareness courses and collaborative projects on environmental and global warming studies are in progress. These initiatives are really helpful in spreading awareness about our contribution to damaging the environment, challenging its sustainability, and driving the concept of *Go Green* for a sustainable environment. Currently, different sectors of our society are using techniques in their operations that belong to green or clean technologies. These are based on the hydrogen and fuel cells technology. A fuel cell is a device that converts the chemical energy from a fuel into electricity by a chemical reaction with an oxidizing agent such as oxygen. Hydrogen is used as a fuel for a fuel cell, although natural gas and some alcohols can also be used as fuels. The basic difference between a fuel cell and a battery is that when the constant supply of fuel and oxygen is interrupted in a battery, or it discharges, then it stops working, whereas in case of a fuel cell, it continues to work until the source of fuel and oxygen is available again. The Welsh physicist William Grove invented the hydrogen fuel cell in 1842. He reversed the process of electrolysis to combine hydrogen with oxygen to generate electricity, separating pure water as a by-product. Also, solid oxide fuel cells are more efficient to power cars when compared with our conventional internal combustion engines, which have an efficiency about 40%–60% of that of cells. Again, we know hydrogen is abundant in our universe. We can get hydrogen from many natural sources, such as natural gas, coal, and so on. But if we are dealing with only green or clean technology, water is the sole source of pollution-free hydrogen. It is used as a primary or secondary source of power generation in many commercial, industrial, and residential buildings, and so on. According to the United Nations, sustainability is defined as "achieving the needs of the present without compromising the ability of future generations to achieve their own needs." Therefore, true sustainability is achieved when everyone, everywhere, can meet their basic needs forever. And sustainable energy means finding clean, renewable sources of energy—sources that renew themselves, rather than sources that can be depleted. There are different forms of renewable energy that can be considered for the sustainable development of a country. The most commonly considered sources are wind, solar, water, bioenergy, and geothermal energy, the details of which have been discussed in the previous section. Therefore, renewable energy can be defined as a category of energy sources that are either directly or indirectly related to the sun, such as solar, hydro energy, and so on. Such sources are known to mankind to be are inexhaustible in nature, such as sunlight, wind, and geothermal heat. And today, this renewable energy is a present need for sustainable urban life. In past years, different agencies across the globe have made efforts to increase the value of these energies among the world's population; thereby, about 16% of world's energy use comes from renewable energy sources, with nearly 10% from biomass and 3.4% from hydroelectricity plants. Today, projects related to renewable energy are of large scale and more suited to urban populations and only a few percent are suited to rural populations, which not only harnesses the potential of renewable energies sources but also help in the sustainable development of mankind. Green industries are those industries that try to minimize their effect on the environment by the implementation of green investment, described as "economics of green growth that is the pathway towards sustainable development by undertaking and implementing the green public investments and public policy initiatives that encourage environmentally responsible private investments." The United Nations Industrial Development Organization (UNIDO) explained that greening of industry is a method to attain sustainable economic growth and promote sustainable economies.

7.3.2 LEDs and Photovoltaics: A Better Alternative in the 21st Century

The use of LEDs and PVs is an ideal way to save energy and the environment if we link two technologies that are related to semiconductors; that is, LEDs and PVs. Due to high development rates, a significant cost reduction has been observed in the past few years. They work in direct current (DC) mode within a compatible voltage range which is suitable and safe to use for household electrical equipment. Also, it is noted that

- PV cells convert light into electricity at a cost of less than 1$/W (installed power, 2017), and the cost of the PV modules is reduced by up to 10% per year.
- Also, the cost of LED modules has dropped below 10$/klm (2017), and the current luminous efficacy of the devices is found to be 130 lm/W.

7.3.3 Main Goals of Development

- *Sustainability*: Meeting the needs of society in ways that can continue indefinitely into the future without damaging or depleting natural resources
- *"Cradle to cradle" design*: Ending the "cradle to grave" cycle of manufactured products by creating products that can be fully reclaimed or reused
- *Source reduction*: Reducing waste and pollution by changing patterns of production and consumption

- *Innovation*: Developing alternatives to technologies
- *Viability*: Creating a center of economic activity around technologies and products that are beneficial to the environment, speeding their implementation, and creating new careers that truly protect our planet

References

1. International Energy Agency, Solar Energy Perspectives: Executive Summary. (2011) Archived from the original (PDF) on 3 December 2011. X. P. Chen, X. Y. Huang, Q. Y. Zhang, *J. Appl. Phys.* 106 (2009) 063518.
2. G. Peter, *Sustainable Energy Systems Engineering: The Complete Green Building Design Resource.* McGraw Hill: New York, NY Professional (ISBN 978-0–07-147359-0) (2007).
3. B. van der Zwaan, A. Rabl, Prospects for PV: a learning curve analysis *Sol. Energy* 74 (2003) 19.
4. O. Morton, *Nature* 443 (2006) 19.
5. T. Boehme, A. R. Wallace, G. P. Harrison, Applying time series to power flow analysis in networks with high wind penetration. *IEEE Trans. Power Syst.* 22 (2007) 951.
6. http://dx.doi.org/10.1109/TPWRS.2007.901610.
7. www.suzlon.com/products.
8. J. W. Lund, Geo-Heat Centre Quarterly Bulletin, Klamath Falls, Oregon: Oregon Institute of Technology, 28 (2007) 1, retrieved April 16, (2009).
9. J. W. Lund, B. Tonya, Geo-Heat Centre Quarterly Bulletin, Klamath Falls, Oregon: Oregon Institute of Technology, 20 (2) (1999) 9–26, retrieved June 2, 2009.
10. S. T. Dye, *Rev. Geophys.* 50(3) (2012) 1.
11. A. Gando, D. A. Dwyer, R. D. McKeown, C. Zhang, *Nat. Geosci.* 4(9), (2011) 647.
12. Geothermal Energy. National Geographic (2009). https://www.nationalgeographic.com/environment/global-warming/geothermal-energy/
13. Light's Labour's Lost: Policies for Energy-Efficient Lighting; 2005 electricity consumption estimated from IEA, World Energy Outlook 2006. Paris, France: (2006).
14. McKinsey & Company, Inc. Lighting the way: Perspectives on the global lighting market. Global McKinsey Quarterly survey, (November 2011).
15. M. Torregrossa, A. Sommer, T. Bagatin. *Guidelines for Sustainable Public Procurement-Led Street Lighting Equipments.* Brussels: Public Procurement Boosts Energy Efficiency (2010).
16. A. D. Almeida, B. Santos, B. Paolo, M. Quicheron, *Renew. Sust. Energy Rev.* 34 (2014) 30.
17. Envirolink Northwest, *Introductory Guide to LED Lighting.* Warrington, UK: Envirolink Northwest (2011).
18. L. Halonen, Energy efficient electric lighting for buildings. In: *Proceedings of IEA Technical Conference,* Copenhagen (2010).

19. M. Möller, J. I. Gröger, A. Weber, I. Wintermayr, Report on the In-depth Identification of Emerging Technologies The SMART SPP consortium, c/o ICLEI–Local Governments for Sustainability, Europe (2009).

20. A.Williams, B. Atkinson, K. Garbesi, F. Rubenstein, E. Page. A Meta-Analysis of Energy Savings from Lighting Controls in Commercial Buildings. Berkeley, CA: Lawrence Berkeley National Laboratory, (2011).

21. IEA, Light's labour's lost policies for energy-efficient lighting (2006).

22. Aalto University, Annex 45—Energy efficient electric lighting for buildings Finland: Aalto University (2010).

23. U.S. Department of Energy (DOE). Light at night: The latest science. Solid-state lighting program. DOE, (2010).

24. L. R. Wilson, B. C. Rowan, N. Robertson, O. Moudam, A. C. Jones, B. S. Richards, *Appl. Opt.* 49 (2010) 1651.

25. K. Barnham, J. L. Marques, J. Hassard, P. O'Brien, *Appl. Phys. Lett.* 76 (2000) 1197.

26. R. Reisfeld, Y. Kalisky, *Nature* 283 (1980) 281.

27. J. Zubia, J. Arrue, *Opt. Fiber Technol.* 7 (2001) 101.

28. W. Yipeng, Black silicon technology to help maintain market dominance of polysilicon. *The tenth Shanghai SNEC Forum Report* (GCL-Poly) (2016).

29. Frank and Qian Yanchao, Directwafer-1366 Photovoltaic Transitional Witness Report (2016).

30. Z. Feng, L. Mengji, Z. Yao, Study on the theory and technology of silicon metal production (II) *Ferroalloys* 40 (2009) 1.

31. Y. Dai, W. Ma, B. Yang, *A Method of Industrial Silicon Purified by Direct Oxidation* (China) (2009) Application No. 200910094078.3[P].

32. M. Ge, C. Cao, J. Huang, S. Li, Z. Chen, K. Zhang, S. Al-Deyab, Y. Lai, *J. Mater. Chem.* A4 (2016) 6772.

33. E. Snyder, T. Watkins, P. Solomon, E. Thoma, R. Williams, G. Hagler, D. Shelow, D. Hindin, V. Kilaru, P. Preuss, *Environ. Sci. Technol.* 47 (2013) 11137.

34. A. Asghar, A. Aziz, A. Mohd, *J. Cleaner Prod.* 87 (2015) 826.

35. N. Eswar, G. Madras, *New J. Chem.* 40 (2016) 3464.

36. Q. Chen, S. Wu, Y. Xin, *Chem. Eng. J.* 302 (2016) 377.

37. M. Imran, A. Yousaf, X. Zhou, K. Liang, Y. Jiang, A. Xu, *Langmuir* 32 (2016) 8980.

38. A. Ajmal , I. Majeed R. Malik H. Idriss M. Nadeem *RSC Adv.* 4 (2014) 37003.

39. D. Quiñones, A. Rey, P. Álvarez, F. Beltrán, G. Puma, *Appl. Catal. B: Environ.* 178 (2015) 74.

40. M. Cruz, C. Gomez, C. Duran-Valle, L. Pastrana-Martínez, J. Faria, A. Silva, M. Faraldos, A. Bahamonde, *Appl. Surf. Sci.* 268 (2015) 1.

41. A. Aboukaïs, E. Abi-Aad, B. Taouk, *Mater. Chem. Phys.* 142 (2013) 564.

42. G. Karaca, Y. Tasdemir, *Sci. Total Environ.* 488 (2014) 356.

43. S. H. S. Chan, T. Yeong Wu, J. C. Juan, C.Y. Teh, *J. Chem. Technol. Biotechnol.* 86 (2011) 1130.

44. H. Sun, S. Liu, S. Liu, S. Wang, *Appl. Catal. B Environ.* 146 (2014) 162.

45. J. Zhang, S. Wageh, A. Al-Ghamdi, J. Yu, *Appl. Catal. B Environ.* 192 (2016) 101.

46. C. Li, Y. Xu, W. Tu G. Chen, R. Xu, *Green Chem.* 19 (2017) 882.

47. W. Ong, L. Tan, Y. H. Ng, S. Yong, S. Chai, *Chem. Rev.* 116 (2016) 7159.

48. D. Masih, Y. Ma, S. Rohani, *Appl. Catal. B Environ.* 206 (2017) 556.

49. X. Wang, S. Blechert, M. Antonietti, *ACS Catal.* 2 (2012) 1596.

50. D. Liaw, K. Wang, Y. Huang, K. Lee, J. Lai, C. Ha, *Prog. Polym. Sci.* 37 (2012) 907.

51. A. Gholami, M. Maddahfar, *J. Mater. Sci. Mater. Electron.* 27 (2016) 6773.
52. Z. Lou, Z. Wang, B. Huang, Y. Dai, *ChemCatChem* 6 (2014) 2456.
53. W. Hou, S. B. Cronin, *Adv. Funct. Mater.* 23 (2013) 1612.
54. M. Xiao, R. Jiang, F. Wang, C. Fang, J. Wang, J. C. Yu, *J. Mater. Chem. A.* 1 (2013) 5790.
55. V. S. Smitha, P. Saju, U. S. Hareesh, G. Swapankumar, K. G. K. Warrier, *Chem. Select.* 1 (2016) 2140.
56. S. Huang, L. Gu, C. Miao, Z. Lou, N. Zhu, H. Yuan, A. Shan, *J. Mater. Chem. A.* 1 (2013) 7874.
57. B. Zheng, Q. Guo, D. Wang, H. Zhang, Y. Zhu, S. Zhou, *J. Am. Ceram. Soc.* 98 (2015) 136.
58. M. Długosz, J. Wa's, K. Szczubiałka, M. Nowakowska, *J. Mater. Chem. A* 2 (2014) 6931.
59. L. D. Palma, E. Petrucci, M. Stoller, *Chem. Eng. Trans.* 60 (2017) (ISBN978-88-95608- 50-1; ISSN2283-9216).

Index